JOURNAL OF CYBER
SECURITY AND MOBILITY

Volume 2, No. 1 (January 2013)

JOURNAL OF CYBER SECURITY AND MOBILITY

Aim
Journal of Cyber Security and Mobility provides an in-depth and holistic view of security and solutions from practical to theoretical aspects. It covers topics that are equally valuable for practitioners as well as those new in the field.

Scope
The journal covers security issues in cyber space and solutions thereof. As cyber space has moved towards the wireless/mobile world, issues in wireless/mobile communications will also be published. The publication will take a holistic view. Some example topics are: security in mobile networks, security and mobility optimization, cyber security, cloud security, Internet of Things (IoT) and machine-to-machine technologies.

JOURNAL OF CYBER SECURITY AND MOBILITY

Volume 2   No. 1   January 2013

Published, sold and distributed by:
River Publishers
P.O. Box 1657
Algade 42
9000 Aalborg
Denmark

Tel.: +45369953197
www.riverpublishers.com

Journal of Cyber Security and Mobility is published four times a year.
Publication programme, 2013: Volume 2 (4 issues)

ISSN 2245-1439 (Print Version)
ISSN 2245-4578 (Online Version)
ISBN 978-87-92982-59-9 (this issue)

Editorial Foreword

It is our pleasure to introduce the readers to the first issue of the second volume of the *Journal of Cyber Security and Mobility*. This is in fact the fourth in the series since the inception of the journal during second World Wide Security Mobility Conference (WWSMC) held at Princeton, NJ in January 2012. In order to provide guaranteed quality of service to the end users one needs to design the security and mobility aspects of the network in an efficient manner. Efficiency of any network is largely determined by the protocols deployed at several layers including physical layer, MAC layer, network layer and application layer. In many cases, a model-based approach is helpful to study the efficiency of these protocols before they are deployed. In some cases, even if the efficient protocols are deployed, quality of service is compromised due to security attacks at the physical layer or network layer, such as denial of service attack caused by jamming at the physical layer or flooding of the network or application layer signaling protocols in the core networks, respectively. Quality of service is also influenced by the topology of the network, access technology and over-the-air link quality. With the advent of heterogeneous access technologies (e.g., CDMA, LTE, GSM), roaming activities may result in interoperability issues thus contributing to the unavailability of the system or inability of a user to make a call when in roaming mode. Thus, it is highly desirable to devise solutions that can provide protection against jamming, provide mechanism for interoperability among different wireless access, devise mechanisms to improve the link quality by extending the range of operation and provide proper authentication mechanism that will secure the physical access to a device.

    The current issue of the journal consists of four papers that address the problem space described above. Each of these papers addresses how the respective problem area affects the end user experience and proposes the solutions that will improve the quality of service by taking care of the respective problem. The first paper by Babar et al., entitled "Activity Modeling and Countermeasure on Jamming Attack", focuses on jamming issues for wireless sensor network (WSN). The authors design a model-based approach that can model behavior of different kinds jamming attacks using activity modeling and can find the counter measures against these jamming attacks. Although they have designed the model for wireless sensor networks (WSN), this model can also be used to study the jamming behavior in other wireless networks such as LTE networks and devise the anti-jamming solutions for other wireless networks. The second paper by Pant et al., entitled 'Intermediate Measurement Node for Extension of WSN Coverage", discusses how the link quality in the radio network affects the quality of service in wireless sensor networks. It takes into account two specific access technologies, namely Wi-Fi and Zig-bee and proposes solutions to increase the extension of WSN coverage. Yaqub et al. describe the physical security problem for the electric vehicles (EV) when these get charged unattended. In their paper, entitled "Prevention of Unauthorized Unplugging of Unattended Recharging EVs", they discuss potential operational problems resulting out of this security breach and propose authentication-based solutions that will prevent unauthorized unplugging of Unattended EVs during recharging. Finally, Arnaud Henry-Labordère in his paper entitled "The Number Continuity Service: Part II – GSM <-> CDMA Seamless Technology Change", highlights the interoperability problem in case of roaming users where a GSM user roams into a CDMA network and vice versa and proposes the roaming solutions by introducing the concepts such as roaming hub that acts as a hub and helps as protocol converter. By means of this solution, one user can maintain seamless interoperability between two different access technologies. Similar concept roaming hub concept can be introduced to take care of interoperability issues among other types of dissimilar technologies such as LTE and UMTS.

The compilation of a journal is a long process and needs help from many people including the authors. We would like to appreciate the help from the reviewers, editorial board members, advisory board members, steering committee members and staff of River Publishers during the review and production process. We hope the readers will find the topics of this series useful and practical. We also solicit contributions from the readers for future isssues of the journal in the area of security and mobility.

Editors-in-Chief

Ashutosh Dutta	*Ruby Lee*	*Neeli Prasad*
AT&T	*Princeton University*	*Aalborg University*

Activity Modelling and Countermeasures on Jamming Attack

Sachin D. Babar, Neeli R. Prasad and Ramjee Prasad

Center for TeleInFrastruktur, Aalborg University, Aalborg, Denmark;
e-mail: {sdb, np, prasad}@es.aau.dk

Received 30 October 2012; Accepted 19 April 2013

Abstract

In last few decades, there has been a wide demand of wireless sensor networks (WSNs) in extensive mission critical applications such as monitoring, industrial control, military, health and many more. The larger demand of WSN in mission critical applications makes it more prone towards malicious users who are trying to invade. These security assaults worsen the performance of WSN in large extent in terms of energy consumption, throughput, and delay. Therefore, it is necessary to save the WSN from these attacks. The major aim of researchers working on WSN security is to enhance the performance of WSN in presence of these attacks. The major concentration of this paper is to model the behaviour of different kinds of jamming attacks using activity modelling and to find countermeasure against it. Jamming attack is the one of the ruinous invasion which blocks the channel by introducing larger amount of noise packets in a network. The activity modelling of jamming attack gives the perfect understanding of its accomplishment on WSN and it is useful to develop the security countermeasure against the attack.

This paper also gives the survey of different countermeasures of WSN and proposes the new countermeasure on jamming attack. The paper suggests the Threshold based Jamming Countermeasure (TJC) on reactive jamming attack which detects the jamming in network and save the network against reactive jamming attack. The implementation of proposed mechanism in different realistic conditions shows that TJC saves the network in case of reactive

Journal of Cyber Security and Mobility, Vol. 2, 1–27.

jamming attack with increased traffic and number of malicious nodes in a network. The paper simulate the TJC by considering the realistic scenarios which shows the adaptability of the algorithm in change traffic interval and mobility among the normal and malicious nodes.

Keywords: Wireless sensor networks (WSNs), activity modelling, security attacks, jamming attacks, media access control (MAC).

1 Introduction

The research in WSN is growing in large perspective to offer the wide variety of application domains. The WSN consist of the large number of nodes which sends the sensed information to the central base station (BS) [1]. The WSN node suffers from large energy constraint because of its limited battery power. The major requirement to achieve quality of service (QoS) in WSN is to reduce energy consumption with minimum delay and maximum throughput. These performance requirements of WSN are largely affected by security attacks which happen at various layers of WSN.

The main objective of this paper is to model the jamming attack [2, 3] which is one of the denials of service attack [4] which blocks the channel by introducing malicious traffic. WSN is vastly invaded by the different kinds of jamming attacks at each layer. The paper mainly concentrates on jamming attacks which occur at physical and medium access control (MAC) layer. Here, it is more effective and destructive because these layers are mainly responsible for allocating the resources. The different kind of active and reactive jamming attack effects on WSN constraints based behaviour, by increasing the energy consumption with increased delay and decreased throughput. These are very important performance parameter for deciding QoS of WSN. The different kinds of jamming attacks are constant jamming, deceptive jamming, random jamming and reactive jamming. All these jamming attacks are modelled to understand the basic sequence of activities during their occurrences in the network. The author uses unified modelling language (UML) [5] based activity modelling approaches for modelling the behaviour of various jamming attacks. Activity modelling models the behaviour by considering different states and shows the various conditions, message transmission between the states. It is one of the useful ways to understand the intelligent behaviour of jamming attack. The activity modelling also gives the understanding of required security solution for reducing the effect of attack on WSN performance.

The second objective is to analyse the different countermeasures on jamming attack. The literature survey shows that most of the solutions on jamming attack are hardware based which are quite expensive to implement and modify. The survey suggests that software based algorithm, which is quite efficient and cost effective way to stop the invasion of jamming attack. The researcher on jamming attack security did a major work for detecting the jamming attack and to reduce the effect of it on QoS of WSN by using some defensive strategies [6]. The defensive strategies can be useful to develop the efficient security model for Internet of Things (IoT) [7]. The task of making defensive strategies will be easier and efficient if we have the full understanding of behaviour of these attacks.

The last objective is to derive the efficient defence mechanism against jamming attack by understanding the behaviour of attacks and different available countermeasures. In this paper we propose a new countermeasure against reactive jamming, namely TJC. The TJC algorithm allows the attack into the network and starts its defensive mechanism once it detects the assaults in a network. It uses threshold based mechanism to detect the attack and to cure it. Here, every node maintains some send threshold value and it compares current transmission with threshold periodically. If it goes beyond that threshold, it understands that an attack has happened and then it applies defensive mechanism. It first detects the jamming node, then informs all neighbouring node about jammer node and change all paths coming from jammed node, i.e. it will put the jammer node out of network. The paper also simulates the TJC algorithm using Network Simulator (NS)-2 by considering realistic conditions. The simulation results show that TJC perform in better manner in existence of reactive jamming attack. It demonstrates good performance of TJC by varying traffic interval and number of malicious nodes in network. The major advantage of TJC is that its defensive mechanism supports with increased number of jamming nodes in a network.

The remaining part of the paper is organized as follows: Section 2 describes the different kinds of jamming attacks with activity modelling of constant jamming, deceptive jamming, random jamming and reactive jamming. Section 3 explains the various countermeasures on jamming detection and prevention. It also surveys widely used defensive mechanisms. Section 4 proposes the jamming countermeasure on reactive jamming attack based on threshold consideration. Section 5 describes the simulation environment for the TJC algorithm and also discusses the various results obtain in presence of reactive jamming. It also shows the comparative discussion of proposed algorithm. Finally, Section 6 concludes the paper with future work.

2 Proposed Activity Modelling of Jamming Attack

The activity modelling explains the functional view of a system by describing or representing logical processes, or functions. Here, each logical process is represented as a sequence of tasks and the decisions that govern when and how they are performed. Activity modelling is one of the UML representations for giving functional view of any processes or tasks [5, 8]. UML is designed to support the description of behaviours that depends upon the results of internal processes. The flow in an activity diagram is driven by the completion of an action. The activity diagram is useful tool to understand the basic flow of security attacks. The next part of this section explains the activity modelling of four different kind of jamming attacks, i.e.

- Constant Jamming Attack
- Deceptive Jamming Attack
- Random Jamming Attack
- Reactive Jamming Attack

2.1 Constant Jamming Attack

Figure 1 shows the activity modelling of constant jamming attack. It gives insight of different activities that takes place during the execution of attack on a network. The sequences of activities are as follows:

- The attacker initiates the constant jamming attack. If attack is successful then node in a network will behave like a constant jammer and start to jam the network, otherwise node will do its regular activity.
- The normal node detects some event and tries to send the data to another node or destination. It checks for availability of channel, if channel is available then it will send data on the channel and send it towards the destination. If channel is not available then it will check for channel repeatedly after some particular interval.
- The jammer node generates the random data after some particular time interval and it will try to send the random data without following MAC rules i.e. without checking for channel.
- The random data generated from the jammer node may collide with data coming from normal node and it jams the whole traffic in the network by increasing the collision in network. The severity of constant jamming will be more if the interval between the random generations of data is too small.

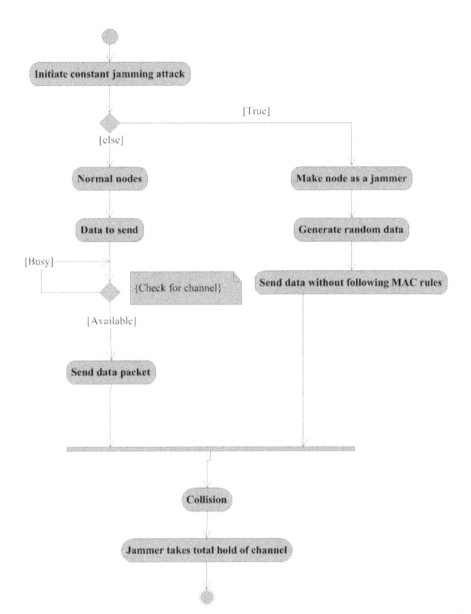

Figure 1 Activity modelling of constant jamming attack.

2.2 Deceptive Jamming Attack

Figure 2 shows the flow of activities in case of deceptive jamming attack. In case of deceptive jamming, attacker will take whole charge of channel by making the channel busy. The different activities that happen during accomplishment of attack are as follows:

- The external attacker initiates the deceptive jamming attack on node in a network. If attack is successful the normal node will act like a deceptive jammer otherwise it will behave like a normal node.
- The normal node generates the data and tries to send the data towards the destination by checking the availability of channel.
- The jammer node generates the data packets continuously without keeping any time gap between the two packets. This continuous generation of packets put the channel in busy state for long time.
- The busy state of channel because of deceptive jamming keeps other normal node to be in receiving state. This behaviour of deceptive jamming increases the energy consumption, delay and decreases the total throughput of the network.

2.3 Random Jamming Attack

Figure 3 shows the different activities that takes place during the execution of random jamming attack. The random jamming attack is kind of intelligent attack where the jamming node thinks for saving of its own energy. Therefore, it works in two modes, i.e. the jamming mode and the sleep mode. The details of execution of attack are as follows:

- If attack is successful, then the external attacker will initiate the attack by converting the normal node into jamming node.
- If channel is available, the normal node detects some event and tries to send the data packet towards another node or destination. The sender node checks for channel availability every time whenever it has data to send.
- The jammer node here works in two modes to save its energy and to last its effect for long time. In jamming mode it make channel busy either by continuously generating packet like deceptive jamming or generate random data after some specified interval without following MAC rules like constant jamming.
- The continuous block of channel by jammer node place the normal node in receive state for long time.

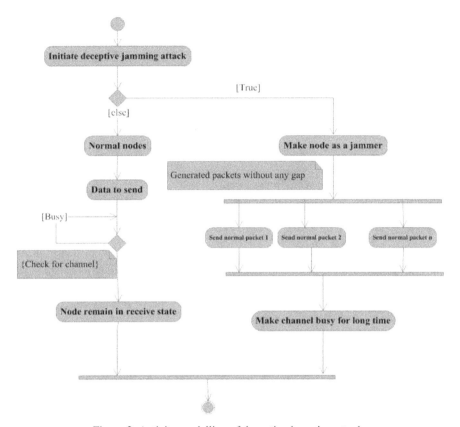

Figure 2 Activity modelling of deceptive jamming attack.

- The normal node changes its receiving state or can get the availability for some time whenever jammer node can go to sleep state. This behaviour of attack introduces the longer amount of delay in the transmission of data from the node.

2.4 Reactive Jamming Attack

Figure 4 shows the activity modelling of reactive jamming. It shows the execution steps of nodes in a network in case of reactive jamming. The steps are as follows:

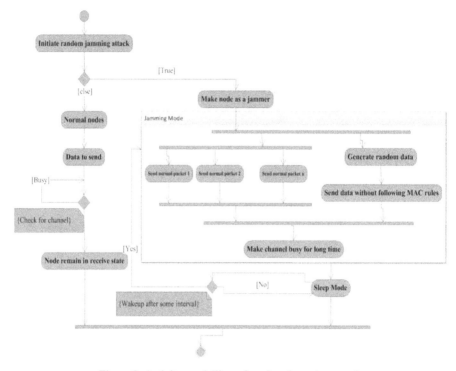

Figure 3 Activity modelling of random jamming attack.

- The reactive jamming attack is initiated by attacking on normal node, if it is successful then node will act like a reactive jammer, otherwise the normal node does its designated operations.
- The main feature of the attack is that it activates when other node in a network are busy to send data or if the channel is busy.
- Here, the normal node tries to send data towards the concern destination by checking the availability of channel and send the data on channel.
- The jammer node check for the channel if channel is ideal it will go to quiet state where it do nothing, else if channel is busy the jammer will activate and generate the noise packet continuously which results in collision in the network.
- The reactive jammer activate when the channel is busy. Therefore, it is very difficult to detect and reduce the effect of channel on performance of network.

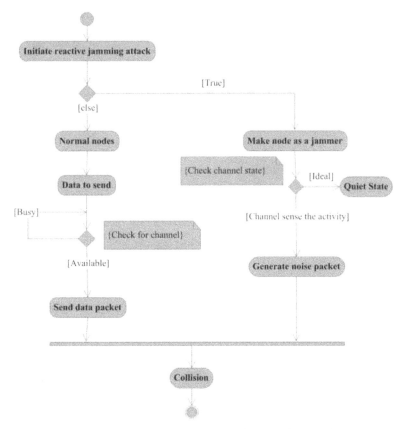

Figure 4 Activity modelling of reactive jamming attack.

3 Related Work on Countermeasures of Jamming Attacks

The security countermeasures against jamming attacks are mainly classified into [2]:

- Detection techniques.
- Proactive countermeasures.
- Reactive countermeasures.
- Mobile agent-based countermeasures.

Detection technique: The purpose of detection technique is to instantly detect jamming attacks. The approaches of these category cannot cope up with jamming alone; they can significantly enhance jamming protection only when used in conjunction with other countermeasures by providing valuable data.

Proactive countermeasures: The role of proactive countermeasures is to make the WSN immune to jamming attacks rather than reactively respond to such incidents. Proactive countermeasures can be classified in software i.e. algorithms for the detection of jamming or encryption of transmitted packets and combined software-hardware countermeasures.

Reactive countermeasures: The main characteristic of reactive counter-measures is that they enable reaction only upon the incident of a jamming attack, sensed by the WSN nodes. Reactive countermeasures can be further classified into software and combined software-hardware.

Mobile-agent based countermeasures: This class of anti-jamming approaches enables Mobile Agents (MAs) to enhance the survivability of WSNs. The term MA refers to an autonomous program with the ability to move from host to host and act on behalf of users towards the completion of an assigned task.

The survey in Table 1 shows the different countermeasures against jamming attack. The table compares all the countermeasures according to the type of technique, mechanism used, its energy efficiency, and implementation cost. The survey gives a varying concluding remark on each kind of countermeasure.

The detection techniques are less efficient according to total energy and implementation cost. Most of the detection technique cannot cope up with jamming attack individually; they require the support of some other countermeasures to work efficiently. The next kind of proactive mechanisms are better than the detection techniques by providing immunity solution to WSN against jamming attack. The proactive countermeasures are mainly classified into proactive software countermeasures and proactive software plus hardware countermeasures. The survey shows that proactive software countermeasure techniques are more efficient than other used techniques because they use some algorithm to defence from jamming instead of allowing the jamming. The proactive countermeasures are efficient solution for active jamming attack such as constant jamming, deceptive jamming and random jamming. The main disadvantage of proactive hardware plus software countermeasure is requirement of hardware, which increases its implementation cost.

The reactive countermeasure technique shows good performance than proactive one in case of reactive jamming attack. Reactive countermeasure allows the jamming in a network and react immediately after the detection of

Table 1 Survey of jamming attack countermeasures.

Countermeasures	Type of technique	Mechanism	Energy efficiency	Implementation cost
The feasibility of launching and detecting jamming attacks in WSNs [9]	Detection technique	It detects the jamming using signal strength or location information.	Low	Low
Radio interference detection protocol (RID) [10]	Detection technique	It uses the interference calculation method and information shared by the node.	Medium	High
Energy-efficient link-layer jamming attacks against WSN MAC protocols [11]	Proactive software	These techniques are mainly embedded inside the MAC to save from jamming effect. The techniques like high duty cycle, shorter data packets, encryption of link layer packet, TDMA protocol, and transmission in randomized interval are used to save from jamming.	Medium	Very low
Defeating energy-efficient jamming [12]	Proactive software	It used frame masking, frequency hopping, and packet fragmentation with redundant encoding.	High	Medium
Hemes II nodes [13]	Proactive hardware and software	It is special kind of node which uses hybrid FHSS-DSSS technique.	Medium	High
A jammed-area mapping service for sensor networks [14]	Reactive software	It detects the jamming by mapping the jam area.	Low	Medium
Channel surfing and spatial retreat [15]	Reactive hardware and software	It uses adaptive channel surfing techniques and spatial retreat mechanism.	High	High
Wormhole-based anti-jamming techniques in sensor networks [16]	Reactive hardware and software	It uses mechanisms like wired pair nodes, frequency hopping pairs with uncoordinated channel hopping.	Medium	High
Jamming attack detection and countermeasures in WSN using ant system [17]	Mobile agent	It used ant algorithm based mobility agent method.	Low	Medium
An algorithm for data fusion and jamming avoidance on WSNs [18]	Mobile agent	It used data fusion mechanism to reduce the effect of jamming and trying to avoid permanently.	Low	Medium
Optimal jamming attacks and network defence policies in wireless sensor networks [19]	Proactive software	Detect the jamming by analysing the percentage of collision and reduce the jamming effect by reducing the collision.	Low	Medium

jamming. They are also classified into reactive software and reactive software plus hardware countermeasures. Here, also reactive software approaches are much cost efficient and energy efficient than reactive hardware plus software countermeasures. The paper mainly concentrates on the software based reactive countermeasure against reactive jamming attack.

The last type of jamming countermeasure is mobile agent based countermeasures. It uses mobile agent who moves host to host to detect the jamming and to do the consigned task of counter measuring against jamming attack. The major disadvantage of this technique is its increase requirement of mobile agent in network, which effects in decreasing efficiency and increase in implementation cost and complexity.

4 Proposed Countermeasure on Reactive Jamming Attack

4.1 Network and Attacker Assumptions

- Network consists of n sensor nodes and one base station (BS).
- All nodes are connected together via bidirectional links.
- The nodes are equipped with synchronized clock, omni-directional antenna and two-ray ground propagation model. Each node is equipped with same capabilities.
- Nodes may communicate directly using single-hop communication or it may communicate using multi-hop communication.
- The nodes are distributed randomly in a network.
- Each sensor node periodically sends a message to the BS.
- The attack can be launch on any node in the network.
- The type of jamming attack assumed is reactive jamming attack, which will be activated when the jammer detects the activity on any node in the network.
- The jammer node is equipped same like a normal sensor node but with capability to generate random jamming signal (random messages).

4.2 Working Mechanism of Proposed Attack

In this paper we propose the threshold based jamming countermeasure (TJC). The key idea of algorithm is to enhance the performance of WSN in presence of reactive jamming attack and to save the WSN from harsh effects of reactive jamming. The algorithm saves the WSN by keeping some threshold at every node. The algorithm achieved it by introducing sending threshold which describe the maximum capabilities of node to send data.

The TJC algorithm works in two phases. The first phase in the threshold based jamming countermeasure is to decide the data sending threshold value of each node. The data sending threshold value is decided at BS side. Here, the BS has capabilities to count and maintain the record of the number of times data send from each node in WSN. Each node is sending the data towards the BS after regular interval, based on amount of data received from particular node per second during normal situation; BS decides the data sending threshold value of each node. BS will maintain the number of average send coming from each node as a sending threshold value.

In the second phase, algorithm will perform the check based on sending threshold value. Here, each node maintains the three states normal state, suspicious state and attacker state. The nodes in normal state are non-attacker node, suspicious state nodes are likely to be an attacker and attacker state nodes are jamming node that started to destroy the network. Initially all nodes are in normal state. The nodes are sending their information to BS either through one-hop or multi-hop way. If the BS is getting more data than expected, i.e. more than the consigned threshold value from the particular source node, then it is changing the state of node as suspicious state. The algorithm will do the path analysis for the suspicious state node; if the suspicious source node is the direct one-hop source then detection of attacker is easy just by doing one-hop path analysis. If the suspicious node is at multi-hop distance from BS then during path analysis phase, algorithm will check for individual node on path for its number of packet transmitted per second. If the number of packets generated by the nodes is more than the average send then that node is considered to be a jammer node and algorithm will make its state as jamming state. Once the jammer node will be detected then algorithm will remove the jammer node outside the path by changing the path through jamming node and also informed to the other neighbouring node to the network that, they have jammer node in neighbour. The detail flow of TJC algorithm is as shown in Figure 5.

5 Simulation of TJC Algorithm and Result Discussion

5.1 Implementation Details

The implementation of all attack is performed by using discrete event simulator NS-2. The parameters set during simulations are shown in Table 2. The idle power, receiving power, transmission power and sleep power are considered according to IEEE 802.15.4 radio model [20].

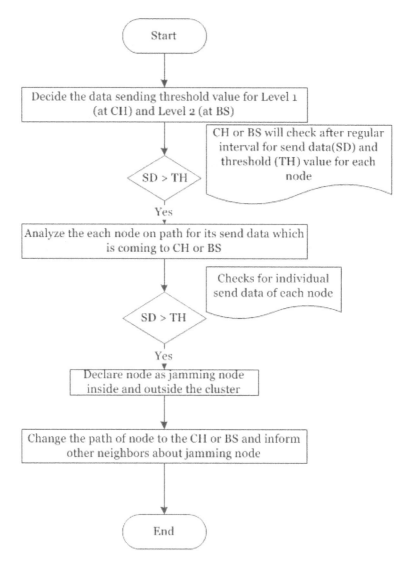

Figure 5 Flow of TJC algorithm.

The simulations are performed in two different conditions. The different conditions are:

- WSN with reactive jamming attack
- WSN with reactive jamming attack with TJC countermeasure

The simulation of jamming attacks is done under following considerations:

- The simulation is performed by varying traffic interval, which is useful to measure the performance of attack and its countermeasures under various traffic conditions. The traffic interval varies from 1 to 10 s. The 1s traffic interval is consider as fast traffic and 10 s traffic interval is consider as slow traffic. These simulations consider number of malicious nodes in network or nodes under attack is one.
- The second set of simulation is performed by varying number of malicious nodes in the network. The number of malicious nodes in network considered is 1, 2, 4, 8 and 16. The traffic interval considers under this simulation is 1 s which is considered to be fast traffic in network. These set of simulations will be useful to analyse the effect of attack and its countermeasures by increasing the destructive entities in a network.
- The third set of simulation is performed by considering some realistic situations where each node is not transmitting information at same time and traffic interval consider is random traffic interval which varies in between 1 to 10 s randomly.
- The last set of simulation is performed by adding random mobility to all nodes in the network. The simulation considers the random traffic interval which varies in between 1 to 10 s randomly. The mobility speed consider here varies from 1 to 25 km/hr. This set of simulations gives the more realistic behaviour of the algorithm by considering random mobility and traffic interval.

5.2 Discussion of Results

5.2.1 Performance by varying traffic interval

Figures 6–8 show the measurement of average energy consumption, delay and throughput by varying the traffic interval respectively. The graphs show that the proposed algorithm TJC improves the energy consumption, delay, and throughput under reactive jamming attack conditions. The algorithm detects the jamming attack by analysing the network and reduces the effect of jamming attack by separating the jamming node from the network.

The energy consumption shown in Figure 6 is less after applying TJC algorithm than in the normal reactive jamming situation. The major reason for enhancing the energy efficiency in TJC is detection of reactive jammer and to place it out of the network. It will help to save the energy consumption happen due to reactive jamming attack.

Table 2 Simulation and node parameters.

Parameter Name	Setting Used
Network interface type	Wireless physical: 802.15.4
Radio propagation model	Two-ray ground
Antenna	Omni-directional antenna
Channel type	Wireless channel
Link layer	Link layer (LL)
Interface queue	Priority queue
Buffer size of IFq	50
MAC	802.15.4
Routing protocol	Ad-hoc routing
Energy model	Energy model
Initial energy (initialEnergy_)	100 J
Idle power (idlePower_)	31 mW
Receiving power (rxPower_)	35 mW
Transmission power (txPower_)	31 mW
Sleep power (sleepPower_)	15 μW
Number of nodes	100
Node placement	Random
Number of simulation runs	50

Figure 7 shows that the delay after applying TJC in a WSN is less than reactive jamming situation because TJC detects the jamming node in network and stop it by keeping it out of the network. The removal of jamming node helps to remove jam on channel, which gives the availability of channel to each node and helps in reduction of delay in case of TJC. In reactive jamming situation, which make channel busy for long time and incur a large waiting time for each node. The busy state of channel also effects on to the throughput of the network, which is improved after applying TJC algorithm as shown in Figure 8.

5.2.2 Performance by Varying Number of Malicious Nodes

Figures 9–11 describe the average energy consumption, delay, and throughput by changing the number of jamming nodes in the network. The number of jamming nodes in network increases from 1 to 16. The figures show that TJC algorithm improves performance against reactive jamming as the number of jamming nodes in network is increasing. The increasing number of jamming nodes in network gives more realistic analysis and adaptivity of TJC if amount of jamming is increasing in the network. The TJC shows efficiency by detecting the multiple jamming on the single path, which shows its perfection to cure the attack.

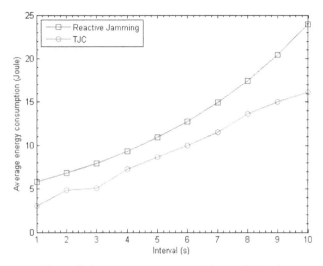

Figure 6 Average energy consumption vs. interval.

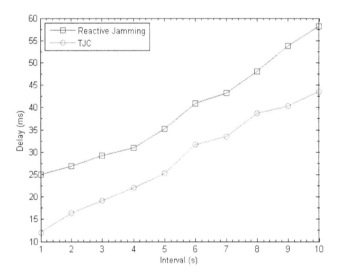

Figure 7 Delay vs. interval.

Figure 9 shows the average energy consumption by varying number of malicious nodes in a network, which shows TJC outperforms as number of malicious nodes is increasing. The major reason of energy saving in case of TJC is its jamming detection mechanism which helps to reduce the en-

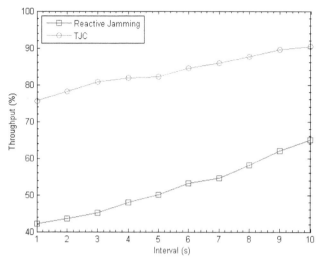

Figure 8 Throughput vs. interval.

ergy consumption due to jamming node and also helps to reduce the energy consumption due to active state of large number of nodes in WSN without sending any data to destination. The detection mechanism of TJC also helps to reduce delay and enhance throughput as shown in Figures 10 and 11. TJC reduce the delay by reducing the channel waiting time and increase through-put by giving quick channel availability to nodes in presence of reactive jamming.

5.2.3 Performance of TJC in Realistic Conditions

Figures 12–14 show the performance of TJC in more realistic situations such as by keeping random interval between the data packets and by transmitting data at different time instead of sending data at same time from each node. The realistic situation gives the more insight picture of performance of TJC in presence of reactive jamming attack. Figure 12 shows the average energy consumption of reactive jamming with and without TJC algorithm by varying number of malicious nodes. It shows that energy efficiency improves after applying TJC in realistic situations too because of technique it uses. The technique used by TJC helps to reduce delay and enhances the throughput as shown in Figures 13 and 14 respectively. The major reason of performance improvement in TJC is because of efficient channel availability compared to reactive jamming.

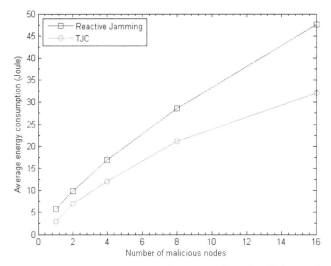

Figure 9 Average energy consumption vs. number of malicious nodes.

5.2.4 Performance of TJC by Considering Mobility

Figures 15–17 show the measurement of average-energy consumption, delay, and throughput respectively by varying the number of malicious nodes in the network. The result shown gives a more truthful support to the presented work because the measurement considers the random mobility among the nodes with random traffic interval. The mobility include in simulation consider the random waypoint mobility model [21]. The mobility scenario helps to check the adaptability of the concern countermeasure in presence of mobility among normal and malicious nodes.

The figures show that as the number of malicious nodes increases in the network, the average-energy consumption and delays also increase with it. The major reason of introducing higher energy consumption and delay is mobility. The mobility among the nodes will take more time to calculate the threshold values for each node, require more energy to scan the path and to detect the location of malicious and neighbouring nodes among it. These reasons lead to increase in energy consumption and delay, they also effects on to the reduction of throughput by increasing the time of jamming detection.

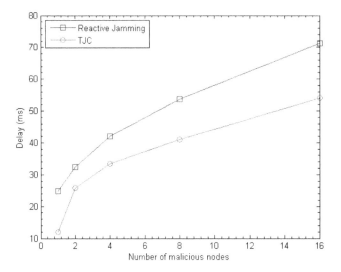

Figure 10 Delay vs. number of malicious nodes.

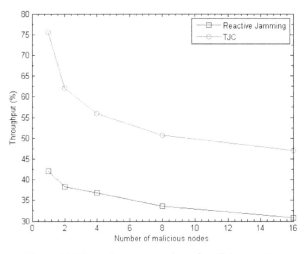

Figure 11 Throughput vs. number of malicious nodes.

6 Conclusions and Future Work

The activity modelling of different jamming attack on WSN provides the functional view of activities executed during accomplishment of the jamming attack. This knowledge is useful tool for the development of efficient security mechanism against jamming attacks. In this paper we also give a survey of

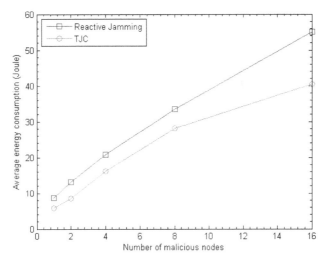

Figure 12 Average energy consumption vs. number of malicious nodes.

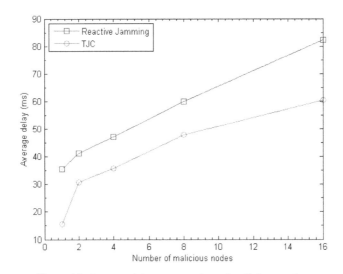

Figure 13 Average delays vs. number of malicious nodes.

existing countermeasures and proposes the countermeasure against reactive jamming attack. The paper proposes TJC countermeasure which shows good performance against reactive jamming attack with varying traffic interval and number of malicious nodes in a network. The proposed TJC algorithm is also tested by considering more realistic conditions where each node is not

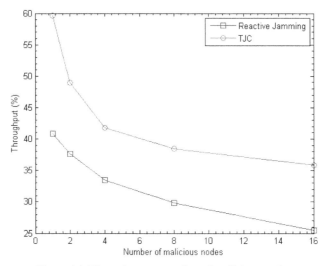

Figure 14 Throughput vs. number of malicious nodes.

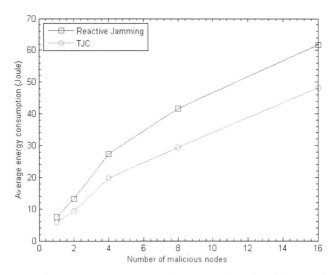

Figure 15 Average energy consumption vs. number of malicious nodes.

transmitting in particular time interval but nodes are transmitting at different time instance. The results under different conditions show that TJC is good solution against reactive jamming attack. The simulation of algorithm by considering mobility shows TJC adaptability with changing position of nodes in the network.

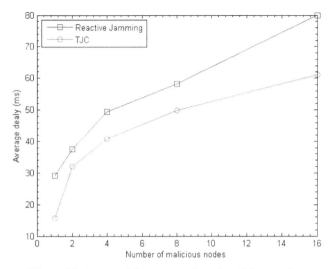

Figure 16 Average delays vs. number of malicious nodes.

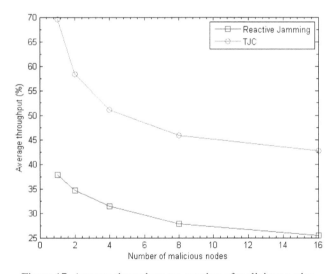

Figure 17 Average throughput vs. number of malicious nodes.

In future research will concentrate on finding a more efficient solution against other kind of jamming attacks by considering the mobility effects. The proposed TJC algorithm can also be extended for cluster based network by distributing task of threshold calculation among the cluster heads (CHs) to save the network from normal reactive jamming and intelligent CH reactive jamming.

References

[1] Jennifer Yick, Biswanath Mukherjee, and Dipak Ghosal. Wireless sensor networks: A survey. Elsevier Computer Networks, 52(12):2292–2330, 2008.
[2] A. Mpitziopoulos, D. Gavalas, C. Konstantopoulos, and G. Pantziou. A survey on jamming attacks and countermeasures in WSNs. IEEE Communications Surveys & Tutorials, 11(4):42–56, 2009.
[3] A.R. Mahmood, H.H. Aly, and M.N. El-Derini. Defending against energy efficient link layer jamming denial of service attack in wireless sensor networks. In Proceedings of IEEE AICCSA, Sharm El-Sheikh, Egypt, 27–30 December, pp. 38–45, 2011.
[4] D.R. Raymond and S.F. Midkiff. Denial-of-service in wireless sensor networks: Attacks and defences. IEEE Journal on Pervasive Computing, 7(1):74–81, 2008.
[5] T. Peder, UML Bible. John Wiley & Sons, 2003.
[6] Wenyuan Xu, Ke Ma, W. Trappe, and Yanyong Zhang. Jamming sensor networks: attack and defense strategies. IEEE Journal on Networks, 20(3):41–47, 2006.
[7] Sachin Babar, Parikshit Mahalle, Antonietta Stango, Neeli Prasad, and Ramjee Prasad. Proposed security model and threat taxonomy for the Internet of Things (IoT). In Springer CNSA, Chennai, India, 23–25 July, pp. 420–429, 2010.
[8] Pranav M. Pawar, Rasmus H. Nielsen, Neeli R. Prasad, Shingo Ohmori, and Ramjee Prasad. Behavioural modelling of WSN MAC layer security attacks: A sequential UML approach. Journal of Cyber Security and Mobility, 1(1):65–82, 2012.
[9] Wenyuan Xu, Wade Trappe, Yanyong Zhang, and Timothy Wood. The feasibility of launching and detecting jamming attacks in wireless networks. In Proceedings ACM MobiHoc, Urbana Champaign, IL, 25–28 May, pp. 46–57, 2005.
[10] G. Zhou, T. He, J.A. Stankovic, and T. Abdelzaher. RID: Radio interference detection in wireless sensor networks. In Proceedings IEEE INFOCOM, Miami, FL, 13–17 March, pp. 891–901, 2005.
[11] Y. Law, L. van Hoesel, J. Doumen, P. Hartel, and P. Havinga. Energy-efficient link-layer jamming attacks against wireless sensor network MAC protocols. ACM Transaction on Sensor Network, 5(1):6.1–6.38, 2009.
[12] A.D. Wood, J.A. Stankovic, and Gang Zhou. DEEJAM: Defeating energy-efficient jamming in IEEE 802.15.4-based wireless networks. In Proceedings IEEE SECON, San Diego, CA, 18–21 June, pp.60–69, 2007.
[13] A. Mpitziopoulos, D. Gavalas, G. Pantziou, and C. Konstantopoulos. Defending wireless sensor nhetworks from jamming attacks. In Proceedings IEEE PIMRC, Athens, Greece, 3–7 September, pp. 1–5, 2007.
[14] A.D. Wood, J.A. Stankovic, and S.H. Son. JAM: A jammed-area mapping service for sensor networks. In Proceedings IEEE RTSS, Cancun, Mexico, 3–5 December, pp. 286–297, 2003.
[15] W. Xu, T. Wood, W. Trappe, and Y. Zhang. Channel surfing and spatial retreats: Defenses against wireless denial of service. In Proceedings ACM Workshop on Wireless Security, New York, 26 September–1 October, pp. 80–89, 2004.

[16] M. Cagalj, S. Capkun, and J.P. Hubaux, Wormhole-based antijamming techniques in sensor networks. IEEE Transactions on Mobile Computing, 6(1):100–114, 2007.

[17] Rajani Muraleedharan and Lisa Osadciw. Jamming attack detection and countermeasures in wireless sensor network using ant system. In Proceedings SPIE, Orlando, FL, 12 March, pp. 1–5, 2006.

[18] A. Mpitziopoulos, D. Gavalas, C. Konstantopoulos, and G. Pantziou. JAID: An algorithm for data fusion and jamming avoidance on distributed sensor networks. Elsevier Journal of Pervasive and Mobile Computing, 5(2):135–147, 2006.

[19] Mingyan Li, I. Koutsopoulos, and R. Poovendran. Optimal jamming attack strategies and network defense policies in wireless sensor networks. IEEE Transactions on Mobile Computing, 9(8):1119–1133, 2010.

[20] Derek J. Corbett, Antonio G. Ruzzelli, David Everitt, and Gregory O'Hare. A procedure for benchmarking MAC protocols used in wireless sensor networks. Technical Report 593, School of IT, University of Sydney, pp. 1–28, August 2006.

[21] C. Bettstetter, G. Resta, and P. Santi. The node distribution of the random waypoint mobility model for wireless ad hoc networks. IEEE Transactions on Mobile Computing, 2(3):257–269, 2003.

Biographies

Sachin D. Babar is ISTE Life Member. He is graduated in Computer Engineering from Pune University, Maharashtra, India in 2002 and received Master in Computer Engineering from Pune University, Maharashtra, India in 2006. From 2002 to 2003, he was working as lecturer in D.Y. Patil College of Engineering, Pune, India. From 2003 to 2004, he was working as lecturer in Bharati Vidyapeeth College of Engineering, Pune, India. From 2005 to 2006, he was working as lecturer in Rajarshi Shahu College of Engineering, Pune, India. From July 2006, he has been working as an Assistant Professor in Department of Information Technology, STES's Sinhgad Institute of Technology, Lonavala, India. Currently he is pursuing his Ph.D. in Wireless Communication at Center for TeleInFrastruktur (CTIF), Aalborg University, Denmark. He has published 20 papers at national and international levels. He has authored two books on subjects like Software Engineering and Analysis of Algorithm & Design. He has received the Cambridge International Certificate for Teachers and Trainers at Professional level under MISSION10X Program. He is IBM DB2 certified professional. His research interests are Data Structures, Algorithms, Theory of Computer Science, IoT and Security.

Neeli Rashmi Prasad, Ph.D., IEEE Senior Member, Director, Center For TeleInfrastructure USA (CTIF-USA), Princeton, USA. She is also Head of Research and Coordinator of Themantic area Network without Borders, Center for TeleInfrastruktur (CTIF) headoffice, Aalborg University, Aalborg, Denmark.

She is leading IoT Testbed at Easy Life Lab (IoT/M2M and eHealth) and Secure Cognito radio network testbed at S-Cogito Lab (Network Management, Security, Planning , etc.). She received her Ph.D. from University of Rome "Tor Vergata", Rome, Italy, in the field of "adaptive security for wireless heterogeneous networks" in 2004 and M.Sc. (Ir.) degree in Electrical Engineering from Delft University of Technology, the Netherlands, in the field of "Indoor Wireless Communications using Slotted ISMA Protocols" in 1997.

She has over 15 years of management and research experience both in industry and academia. She has gained a large and strong experience into the administrative and project coordination of EU-funded and Industrial research projects. She joined Libertel (now Vodafone NL), The Netherlands in 1997. Until May 2001, she worked at Wireless LANs in Wireless Communications and Networking Division of Lucent Technologie, the Netherlands. From June 2001 to July 2003, she was with T-Mobile Netherlands, the Netherlands. Subsequently, from July 2003 to April 2004, at PCOM:I3, Aalborg, Denmark. She has been involved in a number of EU-funded R&D projects, including FP7 CP Betaas for M2M & Cloud, FP7 IP ISISEMD ICt for Demetia, FP7 IP ASPIRE RFID and Middleware, FP7 IP FUTON Wired-Wireless Convergence, FP6 IP eSENSE WSNs, FP6 NoE CRUISE WSNs, FP6 IP MAGNET and FP6 IP Magnet Beyond Secure Personal Networks/Future Internet as the latest ones. She is currently the project coordinator of the FP7 CIP-PSP LIFE 2.0 and IST IP ASPIRE and was project coordinator of FP6 NoE CRUISE. She was also the leader of EC Cluster for Mesh and Sensor Networks and is Counselor of IEEE Student Branch, Aalborg. Her current research interests are in the area of IoT & M2M, Cloud, identity management, mobility and network management; practical radio resource management; security, privacy and trust. Experience in other fields includes physical layer techniques, policy based management, short-range communications. She has published over 160 publications ranging from top journals, international conferences and chapters in books. She is and has been in the organization and TPC member of several international conferences. She is the co-editor is chief of *Journal for Cyber Security and Mobility* by River Publishers and associate editor of *Social Media and Social Networking* by Springer.

Ramjee Prasad (R) is currently the Director of the Center for TeleInfrastruktur (CTIF) at Aalborg University (AAU), Denmark and Professor, Wireless Information Multimedia Communication Chair. He is the Founding Chairman of the Global ICT Standardisation Forum for India (GISFI: www.gisfi.org) established in 2009. GISFI has the purpose of increasing the collaboration between European, Indian, Japanese, North-American, and other worldwide standardization activities in the area of Information and Communication Technology (ICT) and related application areas. He was the Founding Chairman of the HERMES Partnership – a network of leading independent European research centres established in 1997, of which he is now the Honorary Chair.

Ramjee Prasad is the founding editor-in-chief of the Springer *International Journal on Wireless Personal Communications*. He is a member of the editorial board of several other renowned international journals, including those of River Publishers. He is a member of the Steering, Advisory, and Technical Program committees of many renowned annual international conferences, including Wireless Personal Multimedia Communications Symposium (WPMC) and Wireless VITAE. He is a Fellow of the Institute of Electrical and Electronic Engineers (IEEE), USA, the Institution of Electronics and Telecommunications Engineers (IETE), India, the Institution of Engineering and Technology (IET), UK, and a member of the Netherlands Electronics and Radio Society (NERG) and the Danish Engineering Society (IDA). He is also a Knight ("Ridder") of the Order of Dannebrog (2010), a distinguishment awarded by the Queen of Denmark.

Intermediate Measurement Node for Extension of WSN Coverage

Rabin Bilas Pant[1], Hans Petter Halvorsen[1], Frode Skulbru[2]
and Saba Mylvaganam[1]

[1]Faculty of Technology, Department of Electrical, Information Technology and Cybernetics, Department of Energy and Environmental Technology, Telemark University College, Porsgrunn, Norway; e-mail: saba.mylvaganam@hit.no
[2]National Instruments Norge Lensmannslia 4, 1386 Asker, Postboks 177, N-1371 Asker, Norway

Received 15 January 2013; Accepted 3 June 2013

Abstract

Wireless Sensor Networks (WSN) are considered as viable options for data communication for various monitoring and control applications in industries. Improvements in transmission range, security issues, real time monitoring and control issues, system integration and coexistence with other WSN systems are some of the main issues limiting their widespread usage in the industries. After a brief literature survey on some of the recent WSN applications and security management, two designs of WSNs based on Wi-Fi and Zig-bee are presented in this paper. Wireless Distribution System (WDS) and router mode are used in Wi-Fi based and Zig-bee based designs respectively. Zig-bee based design was economic but supports lower sampling rate. Wi-Fi based design was expensive but supports high sampling rate up to 51.2 K samples per second per channel. WSN was setup using NI Zig-bee modules. NI WSN-3202 (sensor node) and NI WSN-9791 (Gateway) were connected in star network topology and multi-hop network topology. Using multi-hop topology, indoor transmission range was increased significantly from 23.1 to 47.1 m with a link quality more than 55%. Since maximum sampling rate of Zig-bee modules is 1 Sample/s per channel, monitoring measurands

Journal of Cyber Security and Mobility, Vol. 2, 29–61.

demanding high sampling rates are deliberately avoided in this work. In both topologies, temperature is the only measuarand handled by both WSN solutions in the solutions presented here.

Keywords: WSN topologies, Wi-Fi, Zig-bee, multi-hop WSN, indoor transmission, sensor node.

Abbreviations

AC	Alternative Current
AES	Advanced Encryption System
AP	Access Point
BAN	Body Area Network
BPSK	Binary Phase Shift Keying
CSS	Central Supervisory Stations
DAQ	Data Acquisition
DO	Dissolved Oxygen
DPSK	Differential Phase Shift Keying
DQPSK	Differential Quadrature Phase Shift Keying
ESF	European Science Foundation
GFSK	Gaussian Frequency Shift Keying
GSM	Global System for Mobile
IEEE	Institute of Electrical and Electronics Engineering
ISM	Industrial, Scientific and Medicine
ISO	International Organization for Standarization
ITS	Intelligent Transportation System
LOS	Line of Sight
MAX	Measurement and Automation eXplorer
MCC	Motor Control Centers
MEMS	Micro-Electro Mechanical system Sensors
NI	National Instruments
NPI	Name Plate Info
OQPSK	Offset Quadrature Phase Shift Keying
OSI	Open System Interconnection
pH	Potential of Hydrogen
RADIUS	Remote Authentication Dial-In User Service
RF	Radio Frequency
S/s or Sa/s	Sampling rate sample/se
SIG	Special Interest Group

TKIP   Temporal Key Integrity Protocol
TUC   Telemark University College
VAC   Voltage Alternating Current
VDC   Voltage Direct Current
WAP   Wireless Access Point
WDS   Wireless Distribution System
Wi-Fi   Wireless Fidelity
WLS   WireLess Sensor
WPA   Wi-Fi Protected Access
WPA2   Wi-Fi Protected Access 2
WSN   Wireless Sensor Network

1 Introduction

A radio frequency (RF) network consisting of sensors, transceivers, machine controllers, microcontrollers and devices facilitating user interfaces is frequently called a Wireless Sensor Network (WSN). Sensors in a WSN are spatially distributed entities monitoring process or environmental parameters such as flow, gas concentration, vibration, pressure, motion, temperature, etc. These parameters will be fused using different sensor fusion strategies at selected nodes or at a dedicated hub in the WSN. WSN has to have at least two nodes communicating with each other. However, in a typical WSN, the numbers of nodes are much than two depending upon the measurement strategies and environments being monitored.

WSN has attracted the attention of both the academia and industries from the late 90s. Some famous prototype smart sensors like Berkeley motes and Smart Dust and solutions based on them have already been implemented [1]. WSN solutions using smart sensors are also commercially available from Crossbow, Philips, Siemens and National Instruments.

Since there are many diverse actors and users involved in the design, development and use of WSN, there is an increasing need for standardization in the field of WSN. For example, National Instrument (NI) WLS-9163 nodes are standardized with Institute of Electrical and Electronics Engineering (IEEE) 802.11 protocols [2] and NI WSN-3202 is standardized with IEEE 802.15.4 protocols [3], as a transmission medium.

WSN has been applied successfully in industries, health and environment sectors. In spite of the advantages in using WSN, two of the major disadvantages are security and coverage area or transmission distance. Since radio communication is used, unauthorized individuals can easily can get access to valuable and sensitive information. A secured network is mandatory in most of the applications today.

There are some important implementation issues in conjunction with the commissioning of any WSN solution. Issues like deployment, mobility, topology, coverage area, life time, sampling rate, cost, energy, etc., have to be looked into when deciding for a WSN based solution for a given problem. Before designing any WSN, the user has to understand its measurement type, its area of coverage, mobility requirement, and budget and then consider design factors like topology, deployment, life time, sampling rate and transmission protocol.

In WSN-application areas like underwater monitoring, bridge structural monitoring, etc., transmission ranges required are much higher than in closed industrial fields, to achieve a strong enough signal link between measuring and monitoring locations.

In most WSN, ISM band frequency is used. One of the limitations of ISM band is the limitation on the output power of antenna 20 dBm [4], the consequence of which is the limits set on coverage areas for the different nodes. WSN using National Instrument Zig-bee devices can provide the highest range of 300 m in America and 150 m in Europe [3] at the cost of reduced sample rate as compared to Wireless Fidelity (Wi-Fi) devices.

Coverage area can be increased by using router node. Router node in WSN is a special type of measurement node which acts as repeaters. These nodes are kept between end node and gateway such that end measurement node first communicates with router node and then gateway. This is a case of multi-hop network. Multi-hop network can be changed to mesh network using more router nodes. A mesh network can provide redundancy in the WSN based network. Another way for coverage extension is the use of multiple Access Point (AP) to create a Wireless Distribution System (WDS). In a WDS system, each AP can communicate with adjacent APs and its associated sensor nodes.

2 Standard Scenario of WSN

Depending on the application areas and the prevailing specific demands on the different properties of WSN, some standardization needs to be strived for, particularly from the point of view of the end-user. Very often, network topology used, bandwidth offered or needed for a particular application, transmission protocol and compatibility of sensors to be integrated in the planned WSN are the main factors to be considered. Figure 1 shows schem-

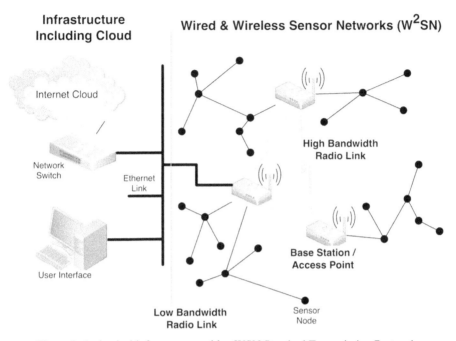

Figure 1 A classical infrastructure with a WSN Standard Transmission Protocol.

atically some of the implications of these demands on WSN in the context of a possible integration of two sub-modules: classical infrastructure and wired and wireless sensor networks.

Classical infrastructure is very often an already existing network structure (predominantly wired with some elements of wireless architecture) with or without internet connectivity. Classical structure may be a simple private or enterprise specific network used for communications and data exchange. Sensor networks as understood in the context of WSN are generally entities spatially distributed, autonomous and connected to each other in an appropriate topology. The different groups of sensors are connected to access points connecting the sensors to end users. End user can monitor and control the process if required. Each access point takes care of the signal traffic and management of signal flow for its group of sensors. In Figure 1, one access point is shown to be connected to end user.

Numbers of sensor nodes and access points differ depending on the type of measurements, environment and coverage area.

Table 1 Zig-bee specifications; Binary Phase Shift Keying (BPSK, Offset Quadrature Phase Shift Keying (OQPSK).

Band (MHz)	Frequency (MHz)	Bit Rate (Kbps)	Symbol Rate (Ksymbol/s)	Modulation
868	868–868.6	20	20	BPSK
915	902–928	40	40	BPSK
2400	2400–2835	250	62.5	O-QPSK

Handling scalar data such as temperature, humidity, pressure, and vibration can be done at low sampling rates, although vibration data will involve relatively higher sampling rates compared to the other parameters. Obviously, image and video data will need still higher sampling rates. As a consequence WSNs are normally used with measurands needing lower sampling rates. The advantages of low scanning rates in a given WSN based measurement are low power and low bandwidth requirements. These are some of the reasons for the deployment of Industrial, Scientific and Medicine (ISM) band in WSN applications. ISM band (2.4 GHZ band) needs no special license and is the choice of Wi-Fi, Zig-bee, Bluetooth and Wireless Hart. As our study uses Zig-bee and Wi-Fi, the following description addresses only these two protocols.

2.1 Zig-bee

Zig-bee is a low cost solution, based on IEEE 802.15.4 standard with ultra-low complexity, and extremely low-power wireless connectivity for portable devices [9]. Important features of Zig-bee are summarized in Table 1.

2.2 Wi-Fi

Wi-Fi is based on IEEE 802.11 protocols operating in ISM bands of 2.4 GHz and 5 MHz [10]. There are five types of Wi-Fi protocols: 802.11, 802.11a, 802.11b, 802.11g, 802.11n. Their ranges vary from 30 to 125 m and data rate varies from 2to 54 Mbps. The newest technology is now 802.11ac (although not commonly available) with even higher data rate. Wi-Fi specifications are summarized in Table 2.

2.3 Standard Network Topology

Arrangement and connection of nodes (circuit elements, sensors, computers, etc.) with each other is called a network topology. A WSN network topo-

Table 2 Wi-Fi specifications.

Types	Range (m)	Bit Rate (Mbps)	Modulation	Band (GHz)
IEEE 802.11	30	1 or 2	FHSS,DSSS	2.4
IEEE 802.11.a	30	54	OFDM	5
IEEE 802.11.b	30	11	DSSS	2.4
IEEE 802.11.g	30	20 to 54	OFDM/DSSS	2.4
IEEE 802.11.n	125	100 to 200	OFDM	2.4
IEEE 802.11ac		500 to 1000	256-QAM	5 GHz

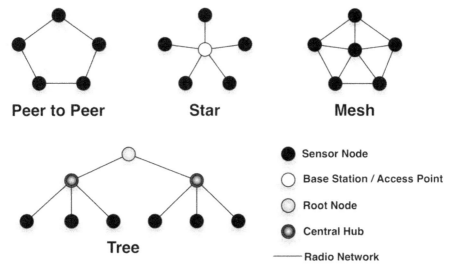

Figure 2 Main network technologies used in WSN applications.

logy indicates how sensor nodes are connected to each other or hubs or base stations. Network topologies include star, ring, fully connected, bus, tree and mesh [12]. Our focus is on peer to peer, star (single point to multipoint), mesh and tree network topologies, which are schematically presented in Figure 2.

In peer to peer topology, each node can communicate directly with another node without any centralized infrastructure or hub.

Remote nodes in a star network can send or receive data to a single base station. In contrast to the operation of peer to peer network, nodes in star network are not permitted to send and receive messages to and from each other. A routing algorithm is simpler in star than those used in other topologies. Some of the disadvantages in star configuration are: mandatory presence of remote nodes within radio range of base station, somewhat reduced redund-

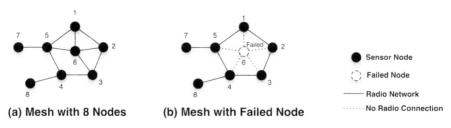

(a) Mesh with 8 Nodes   (b) Mesh with Failed Node

Figure 3 A mesh network with 8 nodes and functionality preserved even with failed nodes.

ancy and robustness. Base stations are central in star topology as they handle messages, routing as well as make the key decisions.

Mesh networks are distributed networks allowing any node in the network to talk to any other node in the same network, of course within the coverage area of the network. Mesh network is appropriate for large scale distributed network of sensors over geographic regions. Personal or vehicle security surveillance systems using mesh networks are described in [12].

Tree network topology is a hybrid of peer to peer network and star network topologies. A tree network topology shown in Figure 2, consists of root node and central hub. Central hub communicates with sensor nodes connected to it and root node is responsible for merging and managing central hub.

However, mesh network topology has the advantages of fault tolerance and load balancing and disadvantages of scalability [13]. As shown in Figure 3, if an individual node fails, a remote node can forward message to desired nodes via any other node in its range.

While deploying a mesh topology one should be very careful to know the supportability of multi-hop structure (node talking to adjacent node) by selected devices. For example, NI-WSN 3202 and Wireless HART support internode communication, whereas, nodes like NI-WLS 9163 can be configured only in star topology.

Advantages of multi-hop network topology are that it helps to increase the distance of communication and adds redundancy feature. It is also true that use of multiple access points can also increase distance in star network topology.

2.4 Extension of WSN Coverage

As described above, the accessible range between source (sensor node) and sink (monitoring node) varies with the transmission protocols used. Zig-bee protocol offers highest transmission range at the cost of lower sampling rate.

Zig-bee devices have coverage up to 300 m. Wi-Fi devices have coverage up to 30 m (802.11ac is the newest, range depending on power levels can approach a couple of km). One obvious method of increasing coverage range is to use higher transmitting power and antennas with more directionally enhanced characteristics. Different countries have varying limits on the maximum output power allowed to be transmitted in ISM bands thus restricting the use of high power directional antennas. Maximum output power allowed in the USA is 50 mW thus extending the maximum range with Zig-bee to 300 m. In Europe and most of Asia the allowed maximum output is 10 mW with an associated maximum range of 150 m [14].

Coverage range can also be extended by adding an additional module in the existing infrastructure between source and sink nodes. This additional module functions as a repeater or router. Access point can be used as an intermediate module in Wi-Fi based WSN and router in Zig-bee based WSN. These two methods are presented in this paper.

2.5 Designing WSN for Coverage Extension

In conjunction with the extension of WSN coverage, the task is to measure any physical quantity at points 300 m away from monitoring stations.

3 Design Alternative Using Wi-Fi

In this design, the concept of WDS is used. WDS is based on the introduction of a new AP between existing sensor node and AP, whereby the overall range of the system is almost doubled. By interlacing the existing WSNs with more APs the wireless distribution is spatially extended. Each intermediary AP acts as a repeater. Many Wi-Fi vendors are using IEEE 802.11a/b/g multimode AP. IEEE 802.11 b/g mode is mainly used to connect sink AP with computer or monitoring station. IEEE 802.11a mode is mainly used to connect two APs in order to form WDS [15]. In the cases presented here, devices supporting IEEE 802.11 b/g multimode are used.

3.1 Technical Specification of Device Used

Devices used in the Wi-Fi design methodology are from National Instruments. NI WLS-9163 C- series carrier along with NI WLS-9234 I/O module are used as sensor nodes. This device supports 4 input channels with 24 bit resolution and maximum sampling rate of 51.2 KS/s per channel using 230

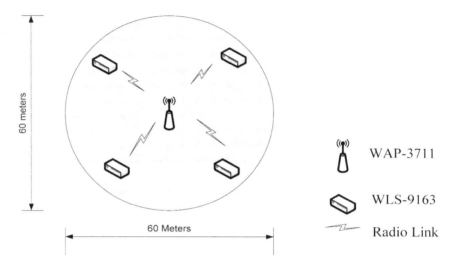

Figure 4 Radiation pattern of single WAP-3711 using omnidirectional antenna.

Volt (VAC) power supply [16]. NI WAP-3701/3711 is used as access point which can support Wi-Fi b/g standard. The range of transmission is 30 m and uses 24 Volt Direct Current (VDC) external power supply [17].

3.2 Proposed Network Architecture

When single WAP-3711 is configured, it can communicate with sensor nodes WLS-9163 up to 30 m. Omnidirectional antenna is used in this module, thus giving a circular radiation pattern of approximately 60 m in diameter, as shown in Figure 4.

WDS with omnidirectional antenna can be advantageously used to extend the coverage range. NI limits the number of WAP-3711 modules to 6 [18]. Each WDS has an overlapping distance of 5 m, such that first WAP- 3711 has range of 55 m and second, third, fourth, fifth have range of 50 m and the last WAP-3711 has range of 25 m. Last WAP-3711 has a range of 25 m because we can use only one side of the circular pattern facing the other modules as shown in Figure 5. This last WAP-3711 is connected to the computer. First WAP-3711 is used to communicate with sensor node WLS-9163 and all the other WAP-3711 modules are used as repeaters. The overall distance of measurement is not more than 330 m in the direction of Line of Sight (LOS).

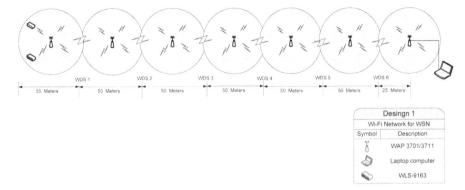

Figure 5 Connection diagram of Wi-Fi WSN using WDS system.

3.3 Design Alternative Using Zig-bee

In this design, the concept of intermediary AP used as repeater is avoided; instead, sensor node is configured as a repeater. An additional sensor node is placed in between the existing sensor node and the gateway. As a result, overall coverage range will be the sum of the coverage ranges of both. This design is based on the concept of multi-hop network. Measuring sensor nodes transfer data to the intermediary sensor nodes, which update and boost the signals and transfer them to the nearest sensor node. In the application presented here, Zig-bee (IEEE 802.15.4) protocol is used and its features are exploited in the design of the system. The advantage of using Zig-bee is mainly in its larger coverage range and cost, compared to the Wi-Fi based solution.

3.4 Technical Specification of Device Used in Design

NI WSN-3202 is used as measurement node and router node. This device supports 4 analog input channels with 16 bit resolution and maximum sampling rate of 1 S/s per channel. It requires 30 VDC power supply when configured as router and standard battery when configured as measurement node. NI WSN-3202 is also facilitated with 4 digital input/output channels [3]. As such, it requires 9 to 30 VDC external power supply. In addition, NI WSN-9791 gateway is used as a sink node, to which the measurement station is connected. The communication range is 300 m in USA and 150 m in Europe. It can support 8 end nodes in star topology and 36 end nodes in mesh topology [14].

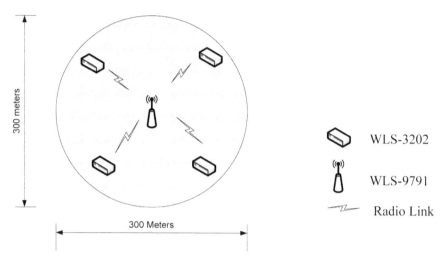

Figure 6 Radiation pattern of single WSN-9791 using omnidirectional antenna.

3.5 Proposed Network Architecture

When a single WSN-9791 is configured, it can communicate with sensor nodes configured as measurement nodes or router nodes. Maximum communication distance is up to 150 m. Because of omnidirectional antenna used in such device, radiation pattern will be a circle of approximately 300 m diameter. Figure 6 shows the radiation pattern of single WSN-9791 using omnidirectional antenna.

The omnidirectional antenna and property of sensor nodes can be used as router nodes to extend the coverage range. To test performance of coverage range extension, two sensor nodes are used thus extending the coverage range to 300 m. One sensor node is configured as router and other as measurement node. We maintain overlapping distance of 5 m between router node and gateway such that WSN-3202 router node has a range of 295 m and WSN-9791 has a range of 145 m. WSN-9791 gateway has a range of 145 m because we can use only one side of its coverage area. A monitoring station or computer is connected as shown in Figure 7, which shows the overall connection of the modules. The overall distance of measurement is not more than 450 m in LOS.

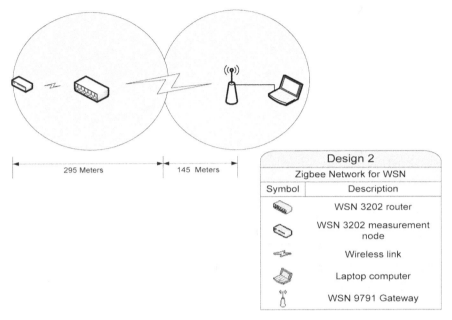

Design 2	
Zigbee Network for WSN	
Symbol	Description
	WSN 3202 router
	WSN 3202 measurement node
	Wireless link
	Laptop computer
	WSN 9791 Gateway

295 Meters   145 Meters

Figure 7 A connection diagram of Zig-bee WSN using sensor node as router.

3.6 Star Network Topology

Full configuration of WSN in a star network topology is given in Figure 8. Both sensor nodes directly communicate with gateway, which in turn, can be monitored by using LabVIEW program in computer. Straight Ethernet cable is used to connect gateway with a laptop computer.

The modules are configured in the LabVIEW platform. After configuring the modules and running the LabVIEW project with them, the system gives rise to the status screen for the sensor nodes and the list of analog and digital signals available in the LabVIEW project as shown in Figures 9 and 10 respectively.

3.7 Determining Maximum Distance of Transmission

After the coupling and configuration of sensor nodes with gateway, second step was to determine maximum wireless distance that WSN devices can support without traffic interruption. For this, sensor node powered with battery was made mobile. The mobile sensor node was slowly taken away from the gateway and the link quality was monitored in LabVIEW program. National

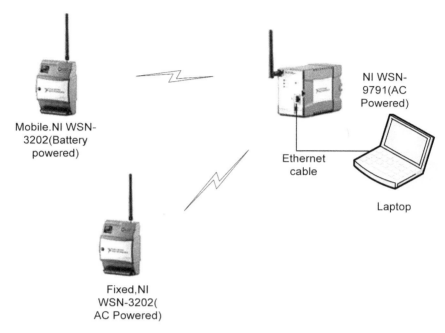

Figure 8 Full configuration in a star network topology using NI modules.

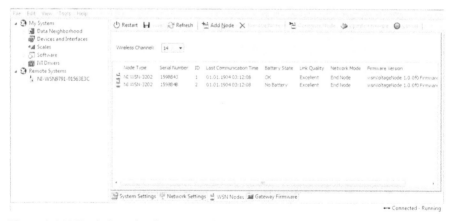

Figure 9 MAX window showing status of both sensor nodes detected by gateway. MAX (Measurement & Automation Explorer) is a graphical user interface provided by NI, to configure IVI (three types of NI drivers). MAX is usually installed with one of the NI application development environments such as LabVIEW or Measurement Studio, or with NI hardware product drivers such as NI-488.2 or NI-DAQ.

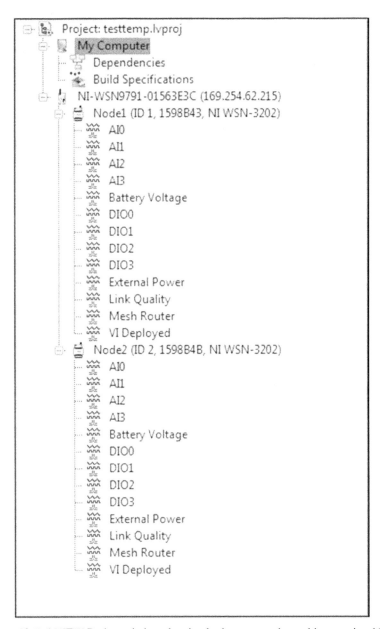

Figure 10 LabVIEW Project window showing both sensor nodes and its associated I/O.

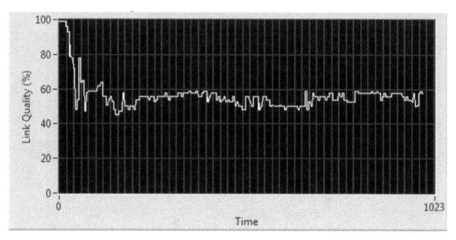

Figure 11 Time vs. link quality graph to determine maximum distance of transmission.

Instruments instructs that 55% of the link quality is supposed to be minimum signal strength and regarded as a fair signal that corresponds to a signal strength of 2 bar highlight in the sensor nodes [39].

A sensor node was moved away further up to the distance where, signal strength decreased to 60%. Measurement of the link quality was done in Lab-VIEW simultaneously. When the signal strength reached to 60%, the sensor node was kept stationary in that place. The distance where the sensor node was kept stationary was measured to be 23.1 m. The link quality fluctuation was measured up to 1024 samples and it was found that most of the time link quality ranges from 55 to 60%. The measurements were based on random selections. Thus, the maximum indoor distance of transmission was determined to be 23.1 m with a link quality from 55 to 60%. Figure 11 shows a LabVIEW plot for a link quality measurement. Up to 168 samples, the sensor node was mobile, then it was fixed in a particular location and other samples were measured.

Sensor node powered with AC was kept stationary at 7.9 m away from the gateway. Figure 12 shows the top view location of both sensor nodes installed and their distance from gateway.

3.8 LabVIEW Programming and Data Interpretation

The program code is presented using the block diagram as shown in Figure 13. The code was written in a while-loop. All the interested I/O variables

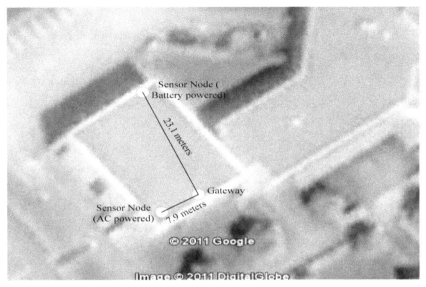

Figure 12 Top view of TUC building to show where the sensor nodes and gateway were installed.

were dragged and dropped to the block diagram form a LabVIEW project window. Indicators for each variable were created. Voltage measurement from sensor nodes was converted to temperature measurement by linear scaling. For signal analysis, Create Histogram Express VI and Statistic Express VI were used. Indicator and Scope for each parameter were created. Shift register and Build array were used to store and process the values in order to calculate arithmetic mean, maximum value, time of maximum value, minimum value, time of minimum value and histogram.

Statistics tab as in Figure 14, consists of scope for different parameters of temperature measurement like, maximum value, minimum value, arithmetic mean, and histogram for both sensors. Numerical values for these measurements along with sample time for maximum and minimum values are also presented.

1024 samples of temperatures were measured. For the temperature variation, window was opened and closed in a room where sensor node 2 (AC powered sensor node) was kept. Sensor node 1 (battery powered sensor node) was kept in big hall, where probability of temperature fluctuations was small.

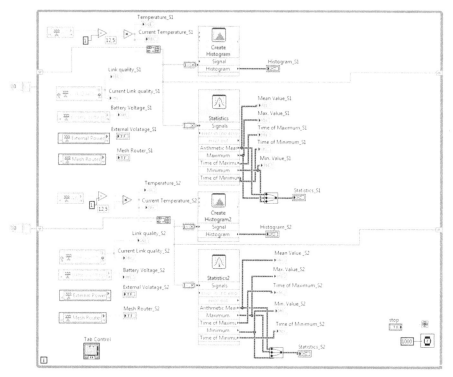

Figure 13 A block diagram of LabVIEW program used to measure temperature from two sensor nodes. NB: This VI is from a set of trial VIs from different student projects and is not an optimized or professional version.

3.9 Multi-hop Network Topology

Sensor node with external power supply was placed stationary at 23.1 m away from the gateway. A distance of 23.1 m was found in the first phase of lab work and this was the maximum distance supported by one sensor node with a link quality of 55% or more. In this phase, stationary sensor node was configured as mesh router mode and battery powered sensor node was configured as end node such that, end node first communicates with intermediate router node and then gateway. Thus, the overall distance of transmission was increased. Full configuration of WSN in multi-hop network topology is given in Figure 16. The gateway is connected to a laptop PC using straight Ethernet cable.

In MAX, sensor node with serial number 159884B was selected, Update Firmware tab was clicked and then from drop down menu, Mesh Router was

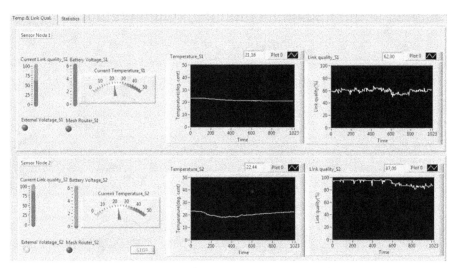

Figure 14 Front panel consisting Temp & Link Qual. tab for both sensors.

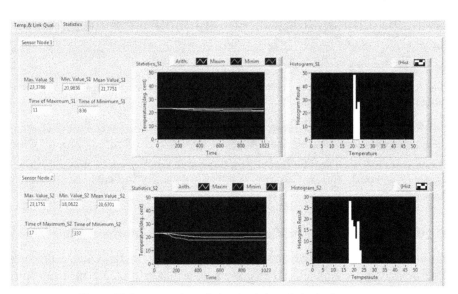

Figure 15 Front panel consisting Statistics tab for both sensors.

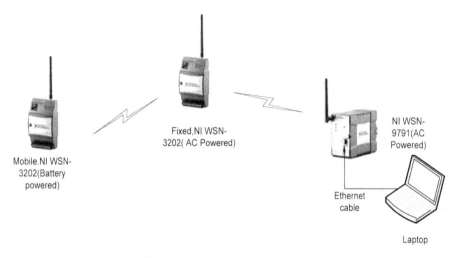

Figure 16 Full configuration in a multi-hop network topology.

Figure 17 MAX window showing Network Mode for both sensor nodes, where one sensor node is updated as Mesh Router.

selected. Sensor node took some time before it was configured to router node. Figure 17 shows the screen shot of MAX screen when one sensor node is updated as router mode. Circle in screen shot clearly indicates Mesh Router mode.

3.10 Determining Operation of Mesh Router Mode

In order to determine the operation of mesh router node, battery powered sensor node was kept closed to router node placed at 23.1 m away from gateway as shown in Figure 18. At first, both nodes communicate directly with gateway with the link quality ranging from 55 to 65%. Reset button of end node was pressed for more than 5 seconds and released. After this,

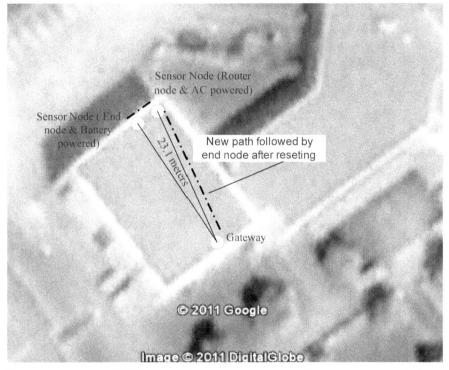

Figure 18 Top view of TUC to show where the sensor nodes and gateway were installed; and new path followed by end node after resetting.

end node searched for the strongest link nearby. Since the nearest link would be the waves propagated by router node, it was connected to router node rather than gateway. The new connection path is shown by the dotted line in Figure 18.

The link quality of end node was measured in LabVIEW throughout the testing period. Figure 19 shows the link quality of end node before and after it has been reset. At first the link quality was seen to be 55–60%. After resetting it was observed that the end node link quality increased sharply and reached almost 100%. This was because of the new link that was established between the end and router nodes.

A significant change in MAX window at this point was observed. The Fair link quality of end node changed to Excellent. This also proves the improvement in link quality after the end node was reset. Figure 20 shows the change in the MAX window when the end node was rested.

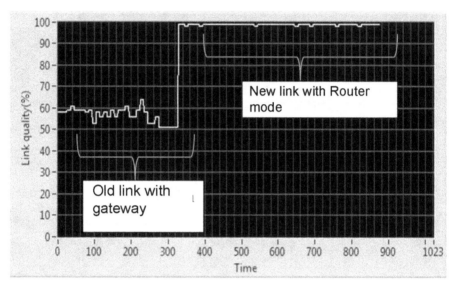

Figure 19 Time vs. link quality graph to observe improvement in link quality after end node has been reset. Note Link quality improvement by almost 50% after 300 s.

Figure 20 MAX window showing 'Excellent' link quality after the end node was reset. It is important to emphasize that this is only the link quality between the node and the router.

Taking advantages of the improved link quality, the end node was moved away from the router node until its link quality reached around 55%. Thus, the overall distance of transmission increased. Figure 21 shows the location of the end and router nodes and their distance from the gateway. The end node was kept 24 m away from router node, thus the total transmission range was measured to be 47.1 m.

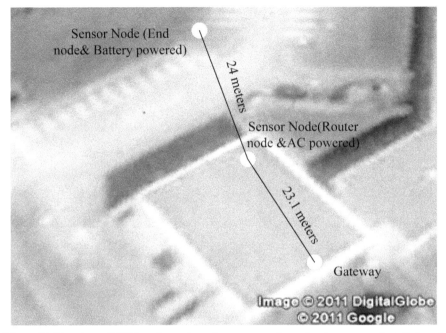

Figure 21 Top view of TUC to show where the sensor nodes and gateway were installed, and their separation distance.

3.11 LabVIEW Programming and Data Interpretation

LabVIEW code used for phase-I of lab work was modified with the provision of measuring temperature statistics for only end node. Since there were no significant variations in temperature, histogram analysis was removed. Code still has provisions for link quality monitoring, battery voltage monitoring along with Boolean indicators for external voltage and mesh router mode for both nodes. Figure 22 shows the code and Figure 23 shows the front panel of LabVIEW program used in this phase of lab work.

4 WSN Implementation Issues

The European Science Foundation (ESF) organized a workshop in April 2004 in order to investigate research in WSN and its practical implications in Europe. Academic researchers and representative form different European country were participated and concluded with important dimensions of the sensor network design [40]. Some of the dimensions of WSN design are

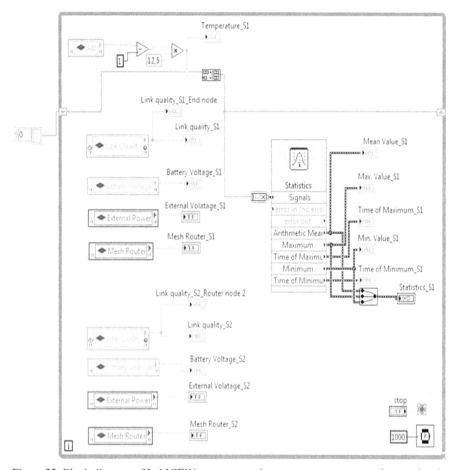

Figure 22 Block diagram of LabVIEW program used to measure temperature from end node.

deployment, mobility, cost, size, energy availability/usage and security. As there are some techniques of handling security with NI modules, this paper focusses on security issues.

4.1 Security

WSN uses RF signal as a physical transmission medium. Wireless medium adds security challenges than that of wired system. Due to this, sensitive data needs to be protected from unauthorized access. There are many common

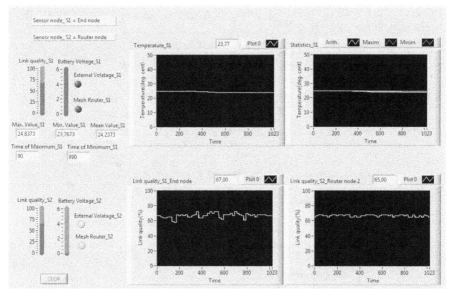

Figure 23 Front panel consisting temperature statistics for end node and link quality statistics for both end node and router node.

security practices, these practices are based on network security components like wireless security protocols, encryption, authentication, etc.

4.2 Security Management

Management of security is done by incorporating appropriate security components as per the nature of networks (public/private, Wi-Fi/Zig-bee). Figure 24 shows the secured WSN architecture. This architecture comprises of WSN, Network Manager and Security Manager. AP or Gateway is connected to Network Manager and Security Manager. Security Manager plays vital role to maintain secured networks by authenticating network devices and by generating, storing and managing encryption keys.

Different transmission protocols have different security types. For example, NI Wi-Fi DAQ 9163 device can supports up to the highest commercially available security IEEE 802.11i known as Wi-Fi Protected Access 2 (WPA2) Enterprises along with IEEE 802.1X authentication and Advanced Encryption Standard (AES) encryption algorithm. However, lower security features like Wired Equivalent Privacy (WEP), Wi-Fi Protected Access (WPA) are also available for this device [44]. The most sensitive Wi-Fi WSN

Infrastructure with Cloud
Including Security Manager

Wired & Wireless Sensor Networks (W^{2}SN)

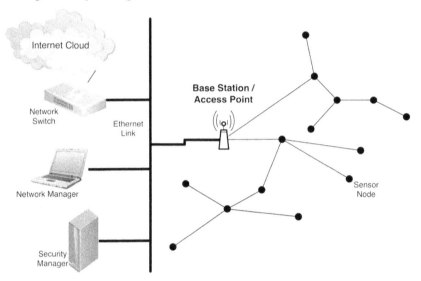

Figure 24 WSN architecture with security components. Adapted with some modifications from [43].

networks in military and industrial application use advanced security management policies with at least one authentication server running a Remote Authentication Dial-In User Service (RADIUS). Less sensitive Wi-Fi WSNs use simple network security; that means that protocols could be WEP or WPA rather than IEEE 802.11i and the encryption key could be Temporal Key Integrity Protocol (TKIP) rather than AES.

5 Design Summaries

- *Protocol and topology*
  Zig-bee seems to be the best protocol if range and power are considered and Wi-Fi seems to be the best protocol if data rate is considered. Zig-bee supports range up to 150 m at the cost of lower data rate of 250 Kbps whereas Wi-Fi supports data rate up to 54 Mbps at the cost of lower range of 30 m. It is important to note that SEA modules for CompactRIO are available for up to 270 meters (inside)0.9 GHz (not for use in EU). The new 802.11ac standard, according to pcmag.com, article 2 of April

9, 2013, supports even up to 1.3 Gbps. Bluetooth has a moderate data rate of 3 Mbps with a maximum range of 100 m.

Mesh network topology have redundant feature. Mesh topology is combination of numbers of multi-hop networks where any node can talk with other nodes in the network. Thus, it is used in such WSN where, nodes are to be distributed in large geographical area. Other network topology like star, tree and point to point cannot give redundant feature and their coverage area are limited to the range of AP.

- *Wi-Fi based vs. Zig-bee based WSNs*

For any transmission range of 300 m, both designs can be used. However, design 1 seems more expensive. The total approximated cost for design 1 implementation is around NOK 119100, whereas the cost for implementing design 2 is much lower, and is around NOK 18100 (all 2011 prices). Design 1 uses Wi-Fi DAQ, it provides high bandwidth and sampling rate of 51.2 K samples per second per channel. However, design 2 uses Zig-bee DAQ, which provides a much lower sampling rate of 1 sample per second per channel. For any physical quantity demanding high sampling rate design 2 fails and for any measurement that requires transmission distance more than 330 m, design 1 fails. This suggest that implementation of WSN in the area which demands both range and sampling rate is practically difficult to design, especially when we consider ISM band and vibration is our interest of measurement.

To tradeoff between range and sampling rate can be solved by modifying our design. Use of National Instruments CompactRIO device along with third party modules [48] can give both high transmission range and high sampling rate. Doing this we can have high sampling rate of Wi-Fi Data Acquisition (DAQ) as design 1, high transmission distance of Zig-bee as in design 2 and new design will be cost effective. For this, NI WLS 9234 can be used as DAQ device and SEA cRIO Zig-bee [49] can be used as transmission modules.

- *Transmission distance*

In star network topology, the maximum distance of communication between sensor node and gateway was found to be 23.1 m with a link quality not lower than 55%. Theoretical transmission distance is 150 m at LOS. But experiment was setup inside the hall, where LOS between sensor nodes and gateway was not possible to maintain. Instruments, machineries placed in hall and closed surface of room where gateway was placed, creates diffraction, reflection, refraction, scattering, shadowing and multipath fading of a signal such that their link quality degrades and overall transmission distance was decreased to 23.1 m rather than 150 m. In order to increase the transmission distance, we need at least two sensor nodes. One sensor node act as router and other node act as end node. Router node acts as repeater that links the end node and gateway. In this mode, the link quality of end node is not dependent on the position of gateway; it entirely depends on position of router node. Hence, end node can be moved away from router node till the link quality between end node and router node becomes 55%. The overall transmission distance was found to be sum of distance from end node to router node and distance form router node to gateway. The total transmission distance was increased from 23.1 to 47.1 m at a fair link quality of 55%. Thus, coverage extension was successfully achieved using router node concept.

- *Security*
  In order to use WSN in application like industrial control and mon-itoring, military purposes, etc., network requires minimum level of security to avoid attacks. Data needs to be sufficiently encrypted, prop-erly authenticated and any change or replay of data during transmission needs to be identified and stopped. Jamming, tampering, collisions, unfairness, misdirection, misinformation, flooding, de-synchronization were identified as different possible attacks when referred to ISO-OSI protocol.
  Jamming and tampering are physical layer attack and can be avoided by spread spectrum technique and tampering free packaging. Data link layer attacks like collision and unfairness can be prevented using col-lision detection techniques and making small frame of data such that they occupy channels for less time. Network layer attacks like packet dropping and misrouting can be prevented by encryption, authentic-ation, multi-path routing and use of unique key. Flooding and de-synchronization are identified as major attacks in transport layer. They are prevented by client puzzle techniques and proper authentication techniques.

6 Conclusion and Future Work

This work is based on experimental work involving predominantly NI hard-ware and software to test performance quality and limitations of intermediate nodes in sensor networking with focus on continuity of the crucial func-tions of the network. The study was done in-campus environments with a lot of structures hindering line of sight communications. However, the basic results indicate the feasibility of achieving improved performance of sensor networks with the help of intermediate nodes. Following the methodology used here might help beginners to learn the basic principles of hardware and software system integration in conjunction with the implementation of intermediate node based sensor networking. After successfully testing the nodes with and without intermediate nodes, the technological possibilities available with NI modules are discussed.

6.1 Conclusion

- Zig-bee provides high range of transmission but less data rate, Wi-Fi provides high data rate but less transmission distance.
- Mesh network topology support multi-hop networking where node can talk to adjacent nodes. This feature allows redundancy in the system. Mesh topology is used when WSN needs to be implemented in large geographical region.
- Wi-Fi based WSN design is expensive as it demands a large number of AP. For the same transmission distance, Zig-bee based design is cheap but suffers less sampling rate problem.
- Tradeoff between sampling rates and transmission distances can be solved by using National Instrument module CompactRIO along with Wi-Fi DAQ and third party Zig-bee transmission module.
- It is useful to see if there are any relationship between time of day and signal strength (disturbance from Wi-Fi phones, radio transmissions, etc.).
- The theoretical range of 150 m was not achieved due to nature of operating environment. LOS cannot be maintained between gateway and sensor node and signal suffers reflection, refraction, scattering, multipath fading, etc.
- Authentication and Encryption are two key security components of wireless networks. Each layer of ISO OSI suffers different types of attack.

6.2 Future Work

- Existing wired system installed in any applications can be replaced by WSN for reliability and performance testing, co-existence problems can be studied by installing Zig-bee based WSN near to any other WSN based in ISM bands.

- Monitoring application can be elaborated to monitoring and control application by integrating WSN with, e.g. a DELTA V process management system.
- Small Network Management Protocol (SNMP) can be implemented in order to access data from any remote computers.
- LabVIEW program can be improved by adding data logging facility, database facility, and alarm handling facility.
- Reliability considerations with data transmission in the 2.4 GHz band.

Acknowledgements

Our thanks are due to Tom-Arne Danielsen of National Instruments of Norway for his valuable comments. We are grateful to Ms. Ru Yan, former PhD research student of TUC for her help with most of the graphics in this paper. Hardware assembly and mechanical work associated work were done by our lab engineers Mr. Eivind Fjelldalen and Mr. Talleiv Skredtvedt. Rabin has now joined the industries in Nepal.

References

[1] P. Santi. Topology Control in Wireless Ad Hoc and Sensor Networks (1st ed.), pp. 9–10. John Wiley & Sons England, 2005.

[2] National Instruments. User guide and specifications NI WLS/ENET-9163. National Instruments Corporations, USA, February 2010.

[3] National Instruments. User guide and specifications NI WSN-3202, National Instruments Corporations, USA, November 2010.

[4] M. Loy, R. Karingattil, and L. Willams. ISM-Band and Short Range Device Regulatory Compliance Overview. Texas Instruments, USA, May 2005.

[5] M.A.E. Villegas, S.Y. Tang, and Y. Qian. Wireless Sensor Network Communication Architecture for Wide-Area Large Scale Soil Moisture Estimation and Wetlands Monitoring, University of Puerto Rico, Puerto Rico, August 2005.

[6] J. Bray and C.F. Sturman. BLUETOOTH 1.1 Connect without Cables (2nd ed.), pp. 2–4. Prentice Hall Inc., USA, 2002.

[7] J.L. Hill. System Architecture for Wireless Sensor Networks. University of California, USA, 2007.

[8] J.G. Castino. Algorithms and Protocols Enhancing Mobility Support for Wireless Sensor Networks Based on Bluetooth and Zigbee. Malardalen University, Sweden, September 2006.

[9] J.A. Gutierrez, E.H. Callaway Jr., and R.L. Barrett Jr. Low-Rate Wireless Personal Area Networks (2nd ed.). IEEE Standards Information Network/IEEE Press, January 2007.

[10] Wi-Fi Alliance. Discover and Learn, 2011, http://www.wi-fi.org/discover_and_learn.php, February 2011.

[11] Javvin. WLAN: Wireless LAN by IEEE 802.11, 802.11a, 802.11b, 802.11g, 802.11n, 2011, http://www.javvin.com/protocolWLAN.html, February 2011.

[12] F.L. Lewis. Wireless Sensor Networks, Smart Environments: Technologies, Protocols and Applications, University of Texas, USA, 2004.

[13] I.F. Akyildiz, X. Wang, and W. Wang. Wireless mesh network: A survey. Computer Networks, 47:445–487, 2005.

[14] National Instruments. User guide and specifications NI WSN-9791 Ethernet Gateway, National Instruments Corporations, USA, November 2010.

[15] J.H. Huang, L.C. Wang, and C.J. Chang. Deployment of Access Point for Outdoor Wireless Local Area Networks, National Chiao Tung University, Taiwan, 2003.

[16] National Instruments. Operating instructions and specifications NI 9234, National Instruments Corporations, USA, August 2008.

[17] National Instruments. NI WAP-3701/3711 User Manual, National Instruments Corporations, USA, September 2007.

[18] National Instruments. Wireless Data Acquisition: Range versus Throughput, National Instruments Corporations, USA, June 2007.

[19] S.Y. Cheung and P. Varaiya. Traffic Surveillance by Wireless Sensor Networks: Final Report, University of California, USA, January 2007.

[20] J. Tavares, F.J. Velez, and J.M. Ferro. Applicaiton of wireless sensor networks to automobiles. Measurement Science Review, 8:65–70, 2008.

[21] L.R. Garcia, L. Lunadei, P. Barreiro, and J.I. Robla. A review of wireless sensor technologies and application in agriculture and food industry: State of the art and current trends. Sensor, 9:4728–4750, 2009.

[22] M. Paavola. Wireless Technologies in Process Automatation – Review and an Application Example. University of Oulu, Finland, December 2007.

[23] Wireless HART. IEC 62591 WirelessHART System Engineering Guide. Emerson Process Management, Revision 2, October 2010.

[24] K. Khakpour and M.H. Shenassa. Industrial Control using Wireless Sensor Networks. K.N. Toosi University of Technology, Iran, 2007.

[25] Department of Energy. Industrial Wireless Efficiency & Renewable Energy Report, USA, December 2002.

[26] J.A. Gutierrez, D.B. Durocher, B. Lu, R.G. Harley, and T.G. Habetler. Applying wireless sensor network in industrial plant energy evaluation and planning systems. In Proceedings of the 2006 IEEE IAS Pulp and Paper Industries Conference, USA, 2006.

[27] M.V. Gangone, M.J. Whelan, and K.D. Janoyan. Deployment of a dense hybrid wireless sensing system for bridge assessment. Structure and Infrastructure Engineering: Maintenance, Management, Life-Cycle Design and Performance, 7:369–378, 2011.

[28] N.D. Battista. Wireless Monitoring the Longitudinal Movement of a Suspension Bridge Deck. University of Sheffield, UK, 2010.

[29] K. Sukun, S. Pakazad, D. Culler, J. Demmel, G. Fenves, S. Glaser, and M. Turon. Health monitoring of civil infrastructures using wireless sensor networks. Information Processing in Sensor Networks, April:254–263, 2007.

[30] S. Kim. Wireless Sensor Networks for Structural Health Monitoring. University of California, USA, 2005.

[31] M. Melo and J. Taveras. Structural Health Monitoring of Golden Gate Bridge Using Wireless Sensor Network – Progress Report. University of Massachusetts Lowel, Aug 2009.

[32] DeltaV. The DeltaV System Overviewed, Emersion Process Management, 2002.

[33] DeltaV. DeltaV OPC.NET Server, Emersion Process Management, January 2011.

[34] DeltaV. Smart Wireless Gateway, Emersion Process Management, 2009.

[35] D. Pompili, T. Melodia, and I.F. Akyildiz. Deployment Analysis in Underwater Acoustic Wireless Sensor Networks. Georgia Institute of Technology, USA, September 2006.

[36] J. Heidemann, Y. Li, A. Sayed, J. Wills, and W. Ye. Underwater Sensor Networking: Research Challenges and Potential Applications. USC/Information Science Institute, USA, 2005.

[37] S. Khodadoustan and M. Hamidzadeh. Tree of Wheels: A New Hierarichal and Scalable Topology for Underwater Sensor Networks. Sharif University of Technology, Iran,2011.

[38] D. Dong. A Survey of Underwater Wireless Sensor Networks – Localization system design. Texas A&M University, USA, 2007.

[39] National Instruments. Why Am I Losing Data When My Node Indicates an Excellent WSN Link Quality?, May 2010, http://digital.ni.com/public.nsf/allkb/C0B57DBB7A7512C0862575EF0058C6C9, March 2011.

[40] K. Romer and F. Mattern. The Desing Space of Wireless Sensor Networks, Institute for Pervasive Computing. ETH, Switzerland, 2004.

[41] A. F. Molish. Wireless Communications, pp. 386–387. John Wiley & Sons Ltd, England, 2005.

[42] M. Chen, T. Kwon, Y. Yuan, and V.C.M. Leung. Mobile agent based wireless sensor networks. Journal of Computers, 1:14–21, April 2006.

[43] H. K. Kalita and A. Kar. Wireless sensor network security analysis. International Journal of Next-Generation Networks, 1:1–10, December 2009.

[44] T. Bakken, R.B. Pant, P. Xie, and A. Shrestha. Wireless Sensor Networks Using NI Modules. Telemark University College, Norway, 2010.

[45] P. Mohanty, S. Panigrahi, N. Sarma, and S.S. Satapathy. Security issues in wireless sensor networks data gathereing protocols: A survey. Journal of Theoretical and Applied Information Technology, 14–26, 2010.

[46] S. Kaplantzis. Security Models for Wireless Sensor Networks. Monash University, Australia, March 2006.

[47] A.D. Wood and J.A. Stankovic. Denial of service in sensor networks. Computer, 35:54–62, October 2002.

[48] National Instruments. CompactRIO Third-Party Modules. National Instruments Corporations, USA, December 2010.

[49] SEA. Zigbee, http://www.sea-gmbh.com/en/products/compactrio-products/sea-crio-Modules/wireless-technology/zigbee/, April 2011.

[50] http://www.broadcom.com/press/release.php?id=s637241, accessed 18 June 2013.

Biographies

Rabin Biplab Pant did his Master degree in the Department of Electrical Engineering, IT and Cybernetics of TUC and has joined the industry in Nepal.

Hans-Petter Halvorsen is Research/Senior Engineer in the Department of Electrical Engineering, IT and Cybernetics of TUC. He works with Research and Development, Programming and System Development, Laboratory Work and Data Acquisition within the fields Measurement and Control Systems. He has worked in industrial IT projects involving LabVIEW and many other programing languages in the Norwegian IT companies CARDIAC and Braze Technology.

Frode Skulbru belongs to the management team of NI in Norway.

Saba Mylvaganam is professor in Process Measurements and Sensorics at TUC.

Prevention of Unauthorized Unplugging of Unattended Recharging EVs

Raziq Yaqub[1], Azzam ul Asar[2], Fahad Butt[2] and Umair Ahmed Qazi[2]

[1]CECOS University, Peshawar, Pakistan; and NIKSUN, Inc., Princeton, NJ, USA;
e-mail: ryaqub@niksun.com
[2]CECOS University, Peshawar, Pakistan; e-mail: fahad_butt44@yahoo.com;
umair.qazi123@hotmail.com

Received 15 January 2013; Accepted 19 April 2013

Abstract

Charging time required by the charging nodes may be up to 5 hours. Due to long charging times, it is possible that drivers will leave their EVs unattended to run other errands while the vehicle is being charged. Unauthorized Unplugging of Unattended (UUU = Triple U, or 3U) vehicle is foreseen as a major issue. The 3U issue will not only result in stoppage of ongoing charging, but also cause frustration for the drivers, and concern for the law enforcer. This paper addresses this problem and presents a method that prevents unauthorized unplugging of EV.

Keywords: Automobile Recharging Nodes (ARN), Unauthorized Unplugging of Unattended Recharging (UUU), electric vehicle.

1 Introduction

Media predict that by 2015 the U.S. will have almost 1 million publicly accessible Electric Vehicle (EV) recharging nodes [1]. The charging time required by the commercial charging facilities may vary from 15 minutes to 5 hours [2] depending on the state of charge of the vehicle, and the Automobile Recharging Node's (ARN) capabilities. Due to long charging times, it is possible that drivers will leave their EVs unattended to run other errands, or kill their time entertaining themselves while the vehicle is being charged. Unauthorized Unplugging of Unattended (UUU = Triple U, or 3U) vehicle is foreseen as a major issue. The 3U issue will not only result in stoppage of ongoing charging, but also cause frustration for the drivers, and concern for the law enforcer. 3U may be committed by misbehaved teens, or EV drivers who might be in a hurry. Such drivers may double park, unplug the charging connector to interrupt ongoing charging, and unethically plug in their own EV for charging. This paper addresses this problem and presents a method that prevents unauthorized unplugging of EV.

According to the method proposed in this paper, upon authentication of the EV driver (e.g. for billing purpose), the ARN will electronically lock the EV charging connector, and establish the security association between the proposed locking mechanism and the EV driver. The security association will result in generating a unique, encrypted, secure digital key (called triple U-Key or 3U-KEY) for the every charging event. The 3U-KEY would be delivered by the ARN to the EV driver upon starting the EV recharging, and would be required to unlock the EV charging connector. Without presentation of the 3U-KEY, the ARN will not unlock the EV charging connector. (The EV charging connector is simply called 'connector' hereinafter.)

The proposed solution will not only prevent Unauthorized Unplugging of Unattended recharging EVs, but also offer several other benefits, e.g. perfect coupling between Charge Coupler (connector and receptacle that connects the electric charging source to electric vehicle), guaranteed disconnection of electrical power supply before attempting to unplug the connector, that would eliminate electrical hazards, capability to unlock the connector by first respondents under potential hazardous or emergency situations even without providing the 3U-KEY, through interaction with the System Administrator.

2 Detailed Description

According to the method proposed here, the ARN is equipped with a mechanism of electronically locking and releasing the charge connector. The electronic lock may exist either in ARN (if the connector and cable is the part of EV, and Receptacle is in the ARN), or in the connector (if the connector and cable is the part of ARN, and Receptacle is in the EV). This mechanism locks the connector in a way that the connector release button either locks or becomes dysfunctional such that the connector cannot be released even on pressing the release connector button. Such locking mechanism locks the connector at the time of charging and delivers an electronic key to the EV driver. The connector is unlocked only by the authorized person who holds the key. Failure in presenting the 3U-KEY to ARN will not unlock the connector.

Electronic key delivery to an EV driver is globally unique for each charging event. We refer to this key as the 3U-KEY. The 3U-KEY is generated by the proposed Unauthorized Unplugging of Unattended EV mechanism, and delivered as encrypted Bar Code, QR Code, or any state of the art secure encrypted code.

The proposed solution will not only prevent Unauthorized Unplugging of Unattended recharging but also offer several other benefits, e.g. (a) perfect coupling between mating parts engaged in high voltage connection, (b) guaranteed disconnection of electrical power supply before unlocking the connector to eliminate electrical hazards, (c) communication capability between system administrator and EV driver/emergency respondents to unlock the connector under hazardous or emergency situations while EV driver is away, or when the system does not accept 3U-KEY due to malfunctioning.

3 Functional Entities

Our method consists of several functional entities as shown in Figure 1. These entities essentially comprise of hardware and software capable of processing tasks pertinent to the present paper. The hardware of the data processing system may comprise of, e.g., lock mechanism, central processing unit (CPU), memory unit, liquid crystal display scanner/printer unit, internal interfaces/buses for wireless/wireline for short/long range communication as shown in Figure 1. The logical entities are discussed in Section 4 and shown in Figure 3.

The whole mechanism works together for authorized disconnection of connector. The 3U-KEY may be:

a. Delivered to the EV driver on a passive device (which does not require a battery, e.g. paper tag). Thus the proposed 3U method has the capability of printing and dispatching the 3U-KEY on a tag. The holder of the passive tag can only unlock and get the released by presenting the physical tag to the ARN. To unlock the connector, the physical possession of the passive tag and the physical presence of tag holder near ARN is required by the system.

b. Delivered to the EV driver on an active device (e.g. mobile device as soft key), thus 3U mechanism has the capability of delivering a 3U-KEY us-

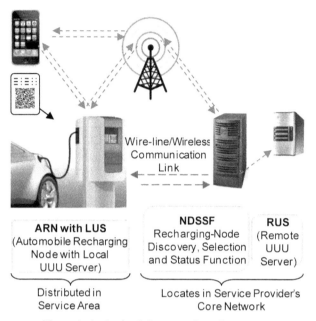

Figure 1   Authorized disconnection of connector.

ing any existing or evolved short range communication, e.g. Near Field Communication. To unlock the connector, the physical presence of the 3U-KEY holder near the ARN is not required.

c. Associated with the EV driver's biometrics. Thus the KGVF memorizes the 3U-KEY, at least until the EV charging event is over and/or the EV is delivered to its driver by presenting the same biometrics. Thus the proposed mechanism has the capability of scanning fingerprint, retina or face, and memorizing the images using any existing or evolved state of the art biometrics technologies. To unlock the connector, the physical presence of the 3U-KEY holder near the ARN is required.

d. Associated with the EV driver's card (e.g. credit card, pre-paid card, gift card, membership card, or any other card or coupon). Thus the KGVF memorizes the 3U-KEY, at least until the EV charging event is over and/or the EV is delivered to its driver by presenting/swiping the same card. Thus the proposed mechanism has the capability of reading cards and memorizing the information using existing or evolved state of the art Card Reading technologies. To unlock the connector, the physical presence of the 3U-KEY holder near the ARN may or may not be required.

e. Delivered using a combination of the above noted schemes to enhance security.

The 3U-Key may also be based on the EV driver credentials, where the ARN may prompt the EV driver to enter his EV license plate number, and a password of his own choice to initiate charging and enters the same to retrieve the EV. This setting has been programmed by students of the CECOS University, Peshawar, Pakistan, and is shown in Figures 2a and 2b.

4 Logical Entities

4.1 Remote 3U Server (RUS)

The RUS is located in the service provider's network. It may be an independent server or a part of Customer Information Server [1]. (In contrast, the LUS is a part of the ARN [1], and is distributed throughout the service area.) RUS and LUS communicate logically through ARN and NDSSF [1] over any available state of the art communication systems, i.e. wireless (LTE, WiMAX), or wireline systems (Optical fiber, or Ethernet, BPL, etc.). Their ability to communicate provides several attractive features, for example:

- Remote recharging status query: The EV driver can send a "recharging status query" to the RUS geographically from anywhere by providing the 3U-KEY in the request message, and will get a response. This feature will be helpful to know the EV recharging status or any other relevant information without physically visiting the location.

- Remote EV claim: The EV driver can deactivate the locking mechanism by providing the 3U-KEY in the message from anywhere to release connector. This feature will be helpful to authorize the EV driver's proxy to claim the EV, without having the EV owner/driver physically visiting the location. Thus the EV may be dropped by one family member and picked up by the other, if authorized by the 3U-KEY holder remotely.

For the above features, the EV driver may send a message, either directly to the RUS only, or the LUS only, or to both, LUS or RUS. Each option has merits and demerits, e.g., the former option offers better security features and system control, whereas the latter reduces system signaling overhead, system messaging efficiency, and response delay, etc. In this paper we propose and support all the options.

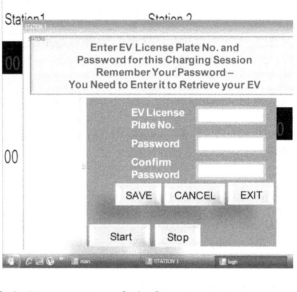

Figure 2 (a) ARN prompts to initiate charging. (b) ARN starts charging.

As shown in Figure 3, the RUS interfaces with a System Administrative. The System Administrative may be a human being or a Voice Activated/Speech Recognition intelligent system. It provides the communication capability between, e.g., Emergency Respondents, or the EV driver and the

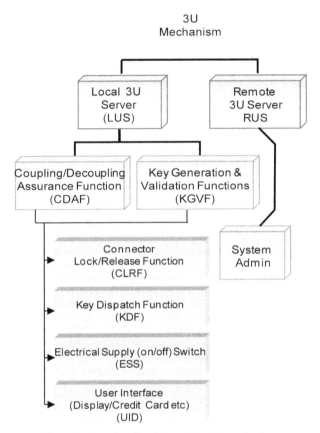

Figure 3 Functional entities of the 3U mechanism.

System Administrator. This feature is useful for several emergency scenarios, for example:

a. the legitimate EV driver lost the 3U-KEY and wants to unlock the connector, or
b. there is potential hazard in the surrounding environment (e.g. a fire) while the EV driver is away and law enforcement agencies need immediate removal of the connector/EV, or
c. the system malfunctions and denies to accept the 3U-KEY, or
d. any other unforeseen reason that mandates to unlock the connector without presenting the 3U-KEY to the system.

4.2 Local 3U Server (LUS)

The LUS is located in the ARN [1]. Upon authentication of the EV driver's credentials (for billing purpose), the LUS will electronically lock the connector. The connector can be unlocked only by presenting the same credentials. Thus the LUS will prevent unauthorized unplugging of unattended EV already being charged.

More specifically according to the method proposed in this paper, upon authentication of the EV driver's credentials the LUS (housed in the ARN) will electronically lock the connector, and establish the security association between the ARN locking mechanism and the EV driver's credentials. The security association will result in generating a unique, encrypted, secure digital key (which we call 3U-KEY) for the every charging event. The 3U-KEY would be delivered by the LUS to the EV driver before starting the EV recharging. The LUS will require the same 3U-Key to unlock the EV charging connector. Without presentation of 3U-KEY, the LUS will not unlock the EV charging connector.

Following the four sub-functional entities assist the LUS in performing its functions. These entities comprise of hardware and, software to perform the tasks pertinent to the present paper. The hardware may, for example, comprise of central processing unit (CPU), memory unit, scanner/printing unit, connector locking mechanism, liquid crystal display (LCD), bus(es) for internal communication, wireless or wireline channels for external communication, and relays/switches, etc. The software may comprise of programs, algorithm and protocols, and so on These functional entities work in coordination:

a. to electronically lock the connector,
b. to establish the security association between the EV driver's credentials and the connector locking mechanism,
c. to generate the 3U-KEY,
d. to deliver the 3U-KEY to the EV driver,
e. to demand the same 3U-KEY to unlock and relinquish the connector, and
f. to unlock the connector on successful presentation and validation of the 3U-KEY.

Unlocking of the connector can be triggered either (a) by the EV driver at any time during or upon completion of the recharging process, or (b) by the LUS upon 100% completion of the charging.

The 3U-KEY may be generated in the form of encrypted Bar Code, QR Code, or any state of the art secure encrypted security Code. The delivery of 3U-KEY to the EV driver may take any one or any combination of the following:

a. Deliver to the EV driver on a passive device (e.g. a tangible/paper tag that does not require a battery). Thus the 3U mechanism has the capability of printing and dispatching the 3U-KEY on a tag using any existing or evolved state of the art mechanisms. The holder of the passive tag can only activate the unlocking mechanism and get the connector released by presenting the physical tag to ARN. Physical possession of the passive tag and physical presence of tag holder near ARN is required by the system.

b. Deliver to the EV driver on an active device (this requires a battery, e.g. mobile communication device as soft key). Thus the 3U mechanism has the capability of delivering a 3U-KEY using any existing or evolved state of the art short range communication, e.g. Near Field Communication (NFC). If the 3U-KEY is delivered to the EV driver's mobile communication device over NFC, the holder of the 3U-KEY can deactivate the lock and get the connector released by presenting the 3U-KEY to ARN in the following two ways:

   i. Presenting the 3U-KEY to the ARN over NFC. NFC is a short-range wireless technology, typically requiring a distance of 4 cm or less. NFC operates at 13.56 MHz and at rates ranging from 106 to 848 kbit/s. NFC always involves an initiator (mobile device in this case) and a target (ARN in this case). Physical possession of the active device and physical presence of the 3U-KEY holder is required to claim the EV.

   ii. Presenting the 3U-KEY to the ARN over a local or wide area network (WLAN, WiMAX, LTE, wireline, or any state of the art network). Physical possession of the active device is required but physical presence of the 3U-KEY holder is not required. Thus the owner of EV may authorize his driver or proxy to claim the vehicle, from the location for which the unlocking step was performed by the owner from remote location.

The EV driver's mobile device may also generate an RF field to read and download the 3U-KEY from a passive tag and deactivate the lock using his mobile device as noted in (a) and (i) above. However, in such a scenario (i.e. LUS issued 3U-KEY on a passive device, and receives 3U-KEY from an active device), additional security measures are built in, e.g. the system may require drivers' license number upon relinquishing the EV.

c. Associate with the EV driver's biometrics. Thus KGVF memorizes the 3U-KEY, at least till the EV charging event is over and/or EV is delivered to its driver by presenting the same biometrics. Thus the proposed mechanism has the capability of scanning fingerprint, retina or face, and memorizing the images using any existing or evolved state of the art biometrics technologies. If the 3U-KEY is associated with the EV driver's Biometrics (based on fingerprint scan, retina scan, or face scan or combination of these) memorized by the ARN, physical presence is required to claim the EV by presenting the same combination of biometrics. (In case of biometrics, it is encouraged implementing combination and or scan of state issued ID card. It would discourage the possibility of dodging the system by using contact lens, gloves, or any other masking means, etc.)

d. Associate with the EV driver's card (e.g. credit card, pre-paid card, gift card, membership card, license, or any other card). Thus KGVF memorizes the 3U-KEY, at least till the EV charging event is over and/or the EV is delivered to its driver by presenting the same card. Thus the proposed mechanism has the capability of reading cards and memorizing the information using existing or evolved state of the art card reading technologies. If the 3U-KEY is associated with the EV driver's Credit Card, memorized by the ARN, physical swipe of the same card is required to claim the EV.

In a charge-status query mode, the 3U-KEY may be presented over local or wide area network (WLAN, WiMAX, LTE, wireline, or any state of the art network) to enquire the status of charging. LUS or RUS, in response, will send the charging status (such as for example fully (100%) charged at time HH:MM:SS on date MM/DD/YYYY, or half (50%) charged at time HH:MM:SS on date MM/DD/YYYY, or any other format most meaningful for the EV driver. (Deactivation of lock from remote may require confirmation to avoid accidental deactivation of the lock, e.g. the user did not intend to unlock the vehicle but just wanted to know the recharging status).

The LUS consists of a Coupling/Decoupling Assurance Function (CDAF) and Key Generation and Validation Functions (KGVF). These are shown in Figure 4.

4.3 Coupling/Decoupling Assurance Function (CDAF)

The CDAF is a major part of LUS and is housed in ARN. It communicates with the RUS (located in the service providers' network), as well as local entities (located with in the ARN), to perform and assure connector lock/release related tasks. Communication with RUS may be over any state of the art long distance wireless or wireline communication system, whereas the communication with local entities may be over any short range communication system (such as wireless personal area networks, wireless local area networks, or any other state of the local wireless or wireline communication link). Before the mechanism unlocks the connector, CDAF makes sure that all the electrical circuits are completely disconnected and high level of shielding and sealing is in place. This offers hazard free disconnect.

4.4 Key (3U-KEY) Generation and Validation Function (KGVF)

KGVF is another major part of LUS and is housed in ARN. It communicates with the RUS (located in the service provider's network) as well as local entities (located with in the ARN) to perform connector lock/release related tasks. The communication with RUS may be over any state of the art long distance wireless or wireline communication system, whereas the communication with local entities may be over any short range communication system (such as wireline or wireless personal area networks, wireless local area networks, or any other state of the communication link).

KGVF is responsible for establishing a security association between the EV driver's credentials and the locking mechanism for any charging event, and consequently generating a 3U-KEY. The 3U-KEY may be generated based on time stamp, recharging node's civic address/GPS coordinates, IP address, ARN identity number, a random key generator, user's credentials (from credit card/membership card, biometrics), any other parameters, and/or any state of the art security key generation algorithm, that ensure a strong security association. (The ARN IP address may be either encrypted or clear text depending on the service provider's business model and security requirements.)

The 3U-KEY can be successfully decoded by any LUS to provide information that may be useful for the EV driver. For example if the EV driver parked the EV for recharging at location "X" and presents the 3U-KEY at location "Y". In such an event, the KGVF may take two actions:

a. Search its own log/database and informs the EV driver that the 3U-KEY was not issued from this location, and

b. Take an additional step i.e. send a query to RUS and get the correct location (civic address) from where the 3U-KEY was issued. This would be beneficial if the EV driver forgot where he parked his car for recharging. However, it may raise security concern, i.e. a stolen/lost-found 3U-KEY may be used to find the EV driver's car, which may not be desirable. Thus this feature may be optionally turned on or off.

c. Take an additional step, i.e. send a query to RUS and get the updated information (e.g. EV is at xx location and yy% charge, or EV had already been claimed at HH:MM:SS on MM/DD/YYYY.

Following local entities assist CDFA and KGVF to achieve the 3U objectives:

- User Interface and Display (UID);
- Connector Lock/Release Function (CLRF);
- Key Dispatch Function (KDF);
- Electrical Supply (On/Off) Switch (ESS).

The functions of CDAF, KGVF, UID, CLRF, KDF and ESS can be easily understood by considering the scenarios presented in Figures 4, 5 and 6.

4.5 3U-KEY Issuance and Recharging Start Event

Figure 4 (with the following description) shows the steps involved in 3U-KEY Issuance and Recharging Start Event:

a = CRR informs CDAF that the connector is coupled

b = UID gets user credentials (e.g. credit card) through user interface and feeds to DKGF. (In this step it also gets user's preference for delivery of the encrypted 3U-KEY, e.g. tag, mobile device, biometrics, card, etc., and activates the related mechanism, e.g. in case of the biometrics option, it activates the scanner for fingerprint/retina/face scan)

c = DKGF authenticates the user with RUS for billing purpose

d = DKGF informs CDAF if the authentication is successful (unsuccessful auth. is informed to UID)

e = CDAF sends message to CLRF to lock the connector

f = CLRF locks the connector and acknowledges back to CDAF

g = CDAF advices On/Off switch to turn on electric supply for EV recharging

h = On/Off switch turns the connection ON and send acknowledgement to CDAF

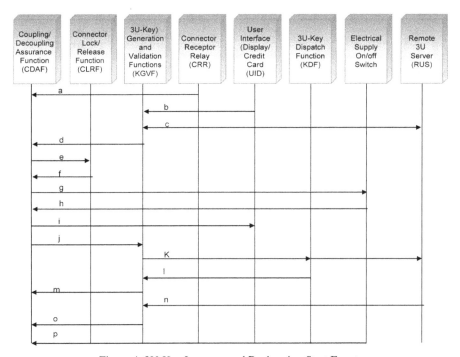

Figure 4 3U-Key Issuance and Recharging Start Event.

i = CDAF informs the user that EV charging started and 3U-KEY generation is in progress

j = CDAF advices DKGF to deliver the 3U-KEY to KDF

k = DKGF delivers the 3U-KEY to KDF (for dispatch to user) and to RUS (for provisioning of value added services and log/record keeping). (KDF delivers the 3U-KEY to the EV driver on the device of his choice, e.g. in case of paper tag, it activates the printer, in case of mobile device, it activates the communication circuitry (e.g. NFC, etc), in case of biometrics or card association, it saves the key in its own internal memory, till the EV charging event is over and driver had claimed the EV by presenting the same card or biometrics)

l = DKGF receives an acknowledgement from KDF

m = CDAF receives Acknowledgement from DKGF

n = DKGF receives an acknowledgement from the RUS

o = CDAF receives acknowledgement from DKGF

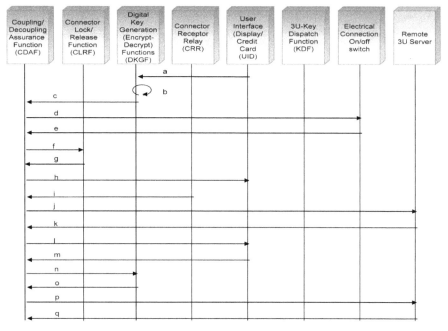

Figure 5 3U-Key Retrieval and Charging Stop Event initiated by the user.

p = ON/Off Switch continuous to measure the status of the battery charging every x seconds and sent to CDAF

If the RUS sends a query to CDAF (e.g. to find recharging status), the CDAF sends a response accordingly (not shown in figure).

4.6 3U-KEY Retrieval and Charging Stop Event Initiated by User

Figure 5 shows the steps involved in a 3U-KEY Retrieval and Charging Stop Event Initiated by the user.

The steps involved in 3U-KEY retrieval and charging stop event initiated by user (per Figure 5) are the following:

a = DKGF receives user's 3U-KEY (e.g. tag, credit card, mobile device) through local UID, (or RUS)

b = DKGF verifies association (decrypts the 3U-KEY) locally and/or with RUS

c = DKGF informs CDAF if the association is verified (unsuccessful verification is informed to UID) d = CDAF sends message to On/Off Switch to turn off electric supply for EV recharging

e = On/Off switch turns the connection OFF and sends acknowledgement to CDAF with energy consumed

f = CDAF sends message to CLRF to unlock the connector

g = CLRF unlocks the connector and acknowledges back to CDAF

h = CDAF informs UID that EV charging is stopped, connector is unlocked and be removed to stop billing

i = CRR informs CDAF that connector is removed

j = CDAF sends RUS the EV charging duration (connector insertion to removal) and energy consumed

k = RUS sends bill (based on charging duration plus energy consumed) to CDAF

l = CDAF sends command to UID to print the receipt

m = UID delivers the receipt to the customer and acknowledges the delivery of receipt to CDAF

n = CDAF advices DKGF to delete the association (RUS retains the information for some predetermined time)

o = DKGF deletes the 3U-KEY Association and send acknowledgement to CDAF

p = CDAF updates RUS the complete log of charging event

q = RUS sends acknowledgement to DKGF (RUS treats the information as specified by the operator)

4.7 3U-KEY Retrieval and Charging Stop Event Initiated by 3U System

Figure 6 shows the steps involved in 3U-KEY Retrieval and Charging Stop Event Initiated by the 3U System.

The steps involved in a 3U-KEY Retrieval and Ccharging Stop Event initiated by the 3U System (per Figure 6) are the following:

a = On/Off Switch turns the connection OFF on 100% charging and sends acknowledgement to CDAF with energy consumed

b = CDAF informs RUS about the 100% completion status

c = RUS sends a text message to the EV driver

d = DKGF receives user's 3U-KEY (e.g. tag, credit card, mobile device) through local UID, or RUS

e = DKGF verifies association (decrypts the 3U-KEY) locally and/or with RUS

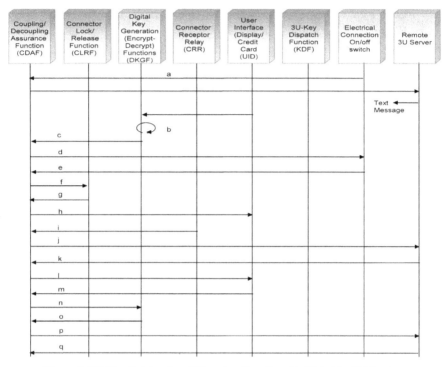

Figure 6 3U-Key Retrieval and Charging Stop Event initiated by 3U System.

f = DKGF informs CDAF if the association is verified (unsuccessful
     verification is informed to UID)
g = CDAF sends message to CLRF to unlock the connector
h = CLRF unlocks the connector and acknowledges back to CDAF
i = CDAF informs the user (UID) that EV charging was stopped at
     hh:mm:ss connector is unlocked and must be removed to stop billing
j = CRR informs CDAF that connector is removed
k = CDAF sends RUS the EV charging duration (between connector
     insertion and removal) and energy consumed
l = RUS sends bill (based on charging duration plus energy consumed)
     to CDAF
m = CDAF sends command to UID to print the receipt
n = UID acknowledges the delivery of receipt
o = CDAF advices DKGF to delete the association
p = DKGF deletes the 3U-KEY Association and sends acknowledgement

to CDAF

q = CDAF advices RUS that charging session is over

r = RUS sends acknowledgment (RUS treats the information as specified by the operator)

The information about any recharging session may be retained by the above noted several components for short term or long term, and may be deleted after some specified time. It may be is application, implementation, or business model specific.

In another embodiment (if the receptacle is in ARN and connector cable is a part of the EV vehicle), the ARN may unlock and automatically release the connector only in the event when 100% charging has been accomplished. This may be an optional provision in the ARN that may be activated or deactivated based on business model. In such a provision the EV manufacturers are proposed to build cable retract feature, i.e. when the ARN releases the connector cable, the EV retracts the cable automatically, and informs the EV driver on the dashboard indicator, or hand held device. This feature would be appreciated by those EV drivers who may prefer to stay inside the car (especially motor-home) during charging, i.e. when the ARN releases the charging cable, the EV retracts the cable automatically, and informs the EV driver. In case of automatic cable retraction provision, the EV manufacturers are proposed to install the cable stow away housing where retracing action does not damage the car body/paint.

The ARN is proposed to include a built-in communication system (comprising microphone, speaker, wired or wires channels, and all the components a modern communication system requires for dealing emergency situations). The built-in communication system would be useful in case of emergencies, such as loss of the 3U-KEY, or any other hazardous conditions wherein it becomes crucial to unplug the EV in the presence or absence of EV driver by law enforcement officer (fire fighter, policeman). Thus the proposed built-in communication system (or independent communication system) will allow the interaction with a System Administrator. The System Administrator would be able to unlock the connector remotely. The System Administrator may use several means (e.g. license number, date of birth, social security number, or any other parameter to verify the communicating person and keep the record).

The user interface utilizes the LCD to interact with the user or to display messages to the user. For example, the LCD may initially display a message, such as "Please insert connector", "How would you like the get the 3U-KEY (printed tag, or over mobile device)", "EV is xx percent charged", "Authentication successful", "Access denied, please try again" or "Please contact the System Administrator", or any other routine messages, error messages, informative message and/or system generated/related message. The display may also have diagnostic capabilities. This feature may be used diagnose the EV's battery health and advertise new battery or auto accessories.

To gain competitive edge, additional services may also be provided by the ARN. The additional services may cover wide range varying from completely technical services (e.g. automobile diagnostic, assessment, etc. and reporting to the EV driver and/or the dealer), to non technical services (e.g. entertainment, or any other service that provides most effective utilization of time while the automobile is being charged. The ARN may also be installed with Automated License Plate Readers (ALPR). This might be appreciated by law enforcement agencies to enhance security and safety of the residents.

References

[1] US Department of Energy. Plug-In Electric Vehicle Handbook for Public Charging Station Hosts. http://www.afdc.energy.gov/pdfs/51227.pdf.
[2] U.S. Department of Energy. Vehicle Technologies Program - Advanced Vehicle Testing Activity Plug-in Hybrid Electric Vehicle Charging Infrastructure Review, Final Report Battelle Energy Alliance Contract No. 58517. http://avt.inl.gov/pdf/phev/phevInfrastructureReport08.pdf.

Biographies

Raziq Yaqub earned a Ph.D. in Wireless Communication from Keio University, Japan, and MBA in Marketing from Fairleigh Dickenson University, USA. He is ere one of the pioneers of LTE/4G, and an inventor of numerous technologies of 4th Generation Wireless Communication, as well as Smart Grid. He received "Innovators Award" from the Governor of the State of New Jersey, USA, through New Jersey Inventors Hall of Fame, for making extra ordinary contributions to the advancement of knowledge and technology.

Dr. Yaqub remained an Executive Director of Toshiba America Research, Inc., from 2001 to 2009, Senior Consultant to the State of New Jersey for 700 MHz LTE Public Safety Network, a spokesperson in 3GPP on behalf of Department of Homeland Security for "Govt. Emergency Telecomm Service", Associate Professor of University of Tennessee in Chattanooga, Adjunct Professor in Stevens Institute of Technology, and now he is Director of Technical Training in NIKSUN, Princeton, USA.

Azzam ul Asar received his Ph.D. and M.Sc. from the University of Strathclyde Glasgow, U.K. He completed his post-doctoral studies from New Jersey Institute of Technology, USA. His research areas include power system, smart grid, microgrid, and intelligent system. He has supervised a number of research projects on energy systems. He is an organizer/session chair of international and national conferences. He has over 100 publications in international journals and conference proceedings. Dr. Asar held several administrative positions during his service including Dean, Chairman, and Director graduate program. Currently, he is full Professor in the Department of Electrical Engineering, CECOS University, Peshawar, Pakistan. He is currently Chair, IEEE Peshawar Subsection, Chair IEEE PES/PEL Joint Society Chapter working under Islamabad Section and an Executive member of IEEE Islamabad Section.

Fahad Butt has done his bachelors degree from University of Engineering and Technology Peshawar in Electrical Power Engineering and his final year project was on control in Smart Grid. Currently he is working in National Engineering Services Pakistan. He is also enrolled in the Masters Programme in CECOS University Peshawar.

Umair Ahmed Qazi did his bachelors from University of Engineering and Technology, UET Peshawar in Electrical Power Engineering and worked in his final year project on power flow control in Smart Grid station. He is presently enrolled in masters program in CECOS University. Currently he is working in a private-sector company named Multinet.

The Number Continuity Service: Part II – GSM $<$-$>$ CDMA Seamless Technology Change

Arnaud Henry-Labordère

HALYS, Paris, France and PRISM-CNRS, Versailles, France; e-mail: ahl@halys.fr

Received 7 May 2013; Accepted 15 May 2013

Abstract

Even with the predominance of the GSM technology, there are instances when the only coverage for a GSM outbound subscriber is from a CDMA (IS-41) network (that includes many fixed networks in the USA and the CDMA based Globalstar satphone services). Symmetrically a CDMA subscriber roaming in a GSM network after changing for a GSM handset, will need the number continuity service to receive his calls and SMS. Roaming Hubs able to convert the two signalling systems are required. We explain the differences into the mobility protocols (MAP GSM and MAP IS-41), the TCAP protocol (ANSI and ITU) and the network layers (14 bits ITU and 24 bits ANSI Point Codes). The eight possible combinations are handled by a multi-standard Roaming Hub architecture. Detailed protocol traces are included.

Keywords: MAP GSM, MAP IS-41, GSM-CDMA converter, ITU, ANSI.

1 History and GSM$<$-$>$CDMA Protocols Comparison

In 2004, a CDMA$<$-$>$GSM number continuity service was provided by Worldcell (USA), mainly for government officials having a CDMA number phone. When they were going abroad, there was no roaming possible,

Journal of Cyber Security and Mobility, Vol. 2, 83–103.

and they had a GSM phone with many roaming agreements rented for this purpose. The "Number continuity" platform developed by Logica, allowed them to receive calls and SMS on their usual US number, and when they were making calls or SMS and their usual US number was appearing as CLI. Since then the platform was sold, but it is not maintained and is then no longer operational. A number continuity project with Globalstar gave a strong reason to redevelop the technology using a more modern Roaming Hub platform, as Globalstar has two types of core networks and terminals (GSM with an Alcatel HLR in Toulouse, and IS-41 (CDMA) with a DSC HLR in Texas.

It is the same system which would allow for example an ordinary CDMA subscriber (example SPRINT in the US) to visit Russia, rent a GSM phone if his own handset is not bi-standard (such as certain Iphones), and get a local IMSI. If this HPLMN has a CDMA<->GSM roaming hub, he would have the full number continuity service. The CDMA<->GSM Hub is still useful as there are many (> 10% of the world mobile users) CDMA networks, in the US (notably SPRINT, Verizon, Metro PCS, Cricket), Asia and Africa (the reason being that the CDMA operators' licences and the core networks are much cheaper). The user changes technology either by getting an other handset or with a mutlti-standard handset (some Iphone versions have the two modes). There exist GSM networks in North America (ANSI) and Europe (ITU), CDMA networks in North America (ANSI) and Europe, Africa or Asia (ITU). Table 1 presents the differences between ANSI and ITU networks whether they are CDMA or GSM.

1.1 TCAP ITU and TCAP ANSI Comparison

However, most GSM networks (T-Mobile USA, AT&T, Canadian GSM) use TCAP ITU even if they are in an ANSI area.

The TCAP ANSI and ITU look similar but are not compatible. It is not just a simple matter of changing the TCAP operation codes, the Component codes and the Transaction Ids also need to be changed. So if interworking needs to be performed between two networks, one with TCAP ANSI, the other TCAP ITU, the Roaming Hub needs two TCAP instances running in parallel.

Table 1 ITU-ANSI differences.

	ITU	ANSI
TCAP layer	TCAP ITU	TCAP ANSI
Transaction codes	Unidirectional = 61 Hex	Unidirectional = E1 Hex
	BEGIN code = 62 Hex	Query with permission = E2 Hex
	CONTINUE = 65 Hex	Continue with permission = E5 Hex
	END = 64 Hex	Response = E4 Hex
	Abort = 67 Hex	Abort 76 Hex
	Does not exist	Query without permission = E3 Hex
	Does not exist	Continue without permission = E6 Hex
Component codes	Invoke = A1 Hex	
	Return result = A2 Hex	
	Return error = A3 Hex	
Origin and Destination Transaction IDs	Reject = A4 Hex	
SCCP layer	SCCP ITU	SCCP ANSI
Network layer: GT formats	Same	Same
Network layer: Point codes	14 bits	24 bits
Network layer	Same (6=HLR, 7=VLR) except	Same (6=HLR, 7=VLR) except
SubSystemNumbers SSN	SMSC GW = 8	SMSC = 11

Table 2 GSM<-> IS-41 differences.

	GSM	IS-41 (used for CDMA)
Name of mobility protocol	MAP GSM (3gpp TS 29.002)	MAP IS-41 (TIA/EIA IS-41 D)
Authentication (VLR<->HLR)	SEND AUTHENTICATION INFORMATION req	AUTHENTICATION req
Registration Circuit Services (VLR<->HLR)	UPDATE LOCATION INSERT SUBSCRIBER DATA	REGISTRATION NOTIFICATION req REGISTRATION NOTIFICATION resp
Incoming call to subscriber's number (GMSC->HLR then HLR->VMSC)	SEND ROUTING INFO req PROVIDE ROAMING NUMBER req	LOCATION REQUEST req ROUTING REQUEST req
Deregistration by user (VLR<->HLR) Incoming SMS to subscriber's number (SMSC->HLR then SMSC->VMSC)	PURGE MS req SEND ROUTING INFO FOR SM req MT FORWARD SM req	MS INACTIVE req SMS REQUEST req SMS POINT TO POINT DELIVERY req
Change of Visited MSC (HLR->old VMSC)	CANCEL LOCATION req	CANCELLATION req
USSD services	PROCESS USSD REQUEST req USSD REQUEST req USSD NOTIFY req	No USSD services in IS-41!!
Data services (Internet)	Circuit mode obsolete, uses ISUP and V110 modems with a IWF GTP protocol (see Section 3)	Circuit mode services only in IS-95. See [10.4] GTP protocol (CDMA packet services in CDMA2000). See [10.4] and [10.5]

Table 2 (Continued)

	GSM	IS-41 (used for CDMA)
Registration Packet Services (VLR <-> HLR)	UPDATE LOCATION GPRS	REGISTRATION NOTIFICATION req ADD SERVICE [10.4] DROP SERVICE [10.5]
Subscriber public number	MSISDN	MDN for outgoing calls, outgoing and incoming SMS DGTSCAR for incoming call (GMSC->HLR)
Mobile Subscriber Identity (in the SIM card (GSM) or in the handset (CDMA))	IMSI	MIN
SMS 7 bits alphabet (text coded in 7 bits is not compatible at all between GSM and IS-41)	3gpp TS 23.038, the 7 bits characters are inside an 8 bits format, with every 8 character filled in the first bit of the 8 bits format	TIA/EAI-637-A, the 7 bits characters are simply packed one after the other in a bit string

1.2 MAP GSM and MAP IS-41 Comparison

Both mobility protocols are called MAP. Table 2 lists the most striking differences. We have given the full list regarding the "number continuity service", including voice and SMS services.

As a consequence of the SSN (Sub System Number) being the same in GSM and IS-41, and of a common international gateway being used by a service provider of number continuity, a routing of the incoming traffic to the MAP GSM stack or the MAP IS-41 stack, cannot be based on the SSN as in most network equipments software (146 goes to Camel, the others to MAP).

For a general operation there must be two routing levels based on a table of Global Titles (GT) specifying the ANSI networks (GSM or CDMA) (so the incoming traffic is sent to TCAP ANSI or the TCAP ITU), and after the TCAP layer, a table specifying the MAP IS-41 or the MAP GSM. A diagram is given in Figure 3.

Such a mixed GSM<->CDMA roaming platform is then much more complex than the implementations which have appeared in the past years, as they necessitate a non standard SS7 architecture using ITU, ANSI, GSM, IS-41 components with some non standard routing levels between the layers. The details given below are for those who want to develop or just need to understand how it works.

To simplify a little we have assumed that the GSM<->CDMA Hub is connected to an ITU SS7 provider which is offering the ANSI<-> ITU Point Code conversion with a partner in the path to the ANSI networks. This is why we see a single MTP3 and M3UA layer as well as a single SCCP ITU layer. If there is no ANSI<->ITU Point Code conversion, it is possible to run with two instances of SCCP and two instances of MTP3 or M3UA.

2 Rerouting of Registration to the GSM<->CDMA Converting Roaming Hub

Figures 5 and 6 of Part I of this article [6] show how the registration messages reach the Roaming Hub that is, SEND AUHENTICATION INFO and UP-DATE LOCATION (GSM HLR handsets), AUTHENTICATON REQUEST and REGISTER NOTIFICATION (CDMA HLR handsets).

As an example, the E164 numbering plan for Globalstar GSM Europe is:

33640044200-44999 for CDMA handsets
33640000000-19999 for GSM handsets

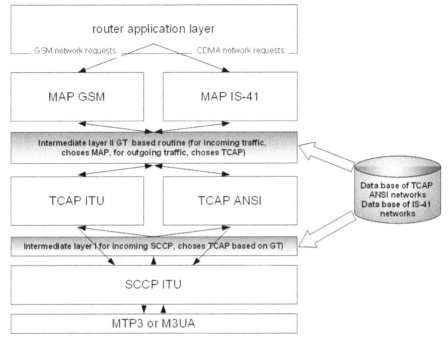

Figure 1 Architecture of a GMS<->CDMA protocol converter.

There are also some ranges for the IMSI (GSM) or MIN (CDMA) assigned to Globalstar Europe.

A sub range of IMSI and MIN is assigned by Globalstar to some planned handsets for the number continuity service. For example:

+208059990040000-49999 for the GSM handsets
+40379810000-40379819999 for the CDMA handsets

and as the subscribers subscribe to the service, their IMSI GSM is entered into the Roaming Hub (IMSI-MIN or IMSI-IMSI mapping) depending on whether they have a CDMA handset or a GSM handset.

In the Gateways GMSCs of Globalstar Europe, a new "proxy" HLR for these ranges of numbers is created, that is an E124 routing table which declares the Roaming Hub as their HLR.

GMSC France +208059990040000-49999 → 33XXXXXXXX (GT of the single roaming Hub)

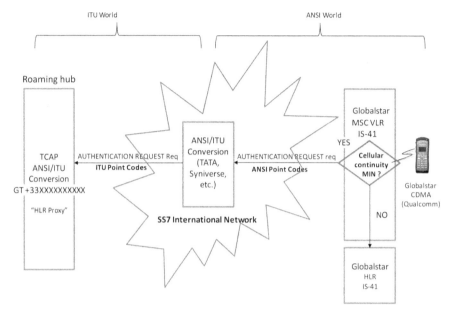

Figure 2 Rerouting to the "HLR proxy".

GMSC USA +40379810000-19999 → 33XXXXXXXX (GT of the single roaming Hub)

This way all the registration messages are forwarded to the Roaming Hub. This is shown in Figure 2 for the rerouting of number continued CDMA Handsets

3 Details of the GSM<->CDMA Number Continuity Implementation

As we assume that most readers are quite familiar with MAP GSM, the protocol analyzer used for the traces displays the equivalence GSM when possible in the IS-41 traces below.

3.1 Authentication VLR<->HLR

The CDMA handset is powered on.

IS-41 Message Decoding © HALYS 2011, 2012
(28): AUTHENTICATION REQUEST (VLR<->HLR Send Authentication Info

in GSM)
Length = 59
    (136): Mobile Identification Number MIN (as IMSI in GSM)
    MIN = +4037990012
    (137): Electronic Serial Number ESN (as IMEI in GSM)
    Manufacturer code = B3
    ESN = B309AFAF
    (149): MSC ID
        7809-10
    (34): System Access Type (SAT)
        (3): Autonomous Registration
    (49): System Capabilities (SYSCAP)
        (0B)
        Authentication parameters were requested on the system access
        Signalling message encryption is supported by the system
        Voice privacy is not supported by the system
        System can execute the CAVE algorithm and share SSD for the
        indicated MS
        SSD is not shared with the system for the indicated MS
    (35): Authentication Response (AUTHR)
        01387C
    (10): Count Update Record (COUNT)
        (00)
    (32): PC_SSN
        PC (24 bits) = 2247429 SSN = 7
    (40): Random Variable (RAND)
        08016558
    (47): Terminal Type (TERMTYPE)
        (32): IS-95

3.2 Registration

After a successful answer from the HLR, the subscriber registers and the profile send by the HLR is loaded in the VLR.

3.2.1 VLR->HLR Request

IS-41 Message Decoding © HALYS 2011, 2012
(13): REGISTRATION NOTIFICATION (VLR-¿HLR Update Location or HLR->VLR Insert Subscriber Data in GSM)
Length = 59

4097 (136): Mobile Identification Number MIN (as IMSI in GSM)
MIN = +4037990012
(137): Electronic Serial Number ESN (as IMEI in GSM)
Manufacturer code = B3
ESN = B309AFAF
(145): Qualification Information code (QUALCODE)
(3): Validation and profile
(150): System My Type Code (MYTYPE)
(16): QUALCOMM
(149): MSC ID
7809-10
(32): PC_SSN
PC (24 bits) = 2247429 SSN = 7
(104): SMS Address (as Visited MSC GT in GSM)
Type of digit 00
Nature of number 01
International
Presentation allowed
Number is not available
(2): Telephony Numbering E164
(1): BCD
Number of digits 11
+16139889998
(53): Extended MSC Identification Number (EXTMSCID)
7809-200
(49): System Capabilities (SYSCAP)
(0B)
Authentication parameters were requested on the system access
Signalling message encryption is supported by the system
Voice privacy is not supported by the system
System can execute the CAVE algorithm and share SSD for the indicated MS
SSD is not shared with the system for the indicated MS

3.2.2 HLR->VLR Response with the Subscriber's Profile

IS-41 Message Decoding © HALYS 2011, 2012
(13): REGISTRATION NOTIFICATION (VLR->HLR Update Location or
HLR->VLR Insert Subscriber Data in GSM)
Length = 65
(150): System My Type Code (MYTYPE)
(63): Globalstar
(142): Authorization Period (AUTHPER)

(6): Indefinite,value = 0

(149): MSC ID

7808-222

(78): Authentication Capabilities

(1): No authentication required

(153): Calling Features Indicator (as Call Forwarding Conditions in GSM) (CFI)

CFNA CFB CFU CD CNIR CNIP1

(93): Mobile Directory Number MDN (as MSISDN for the SRI_FOR_SM or in the INSERT SUBSCRIBER DATA GSM)

Type of digit 05

Nature of number 31

International

Presentation allowed

Number is not available

(2): Telephony Numbering E164

(1): BCD

Number of digits 10

MDN = +4037990012

(151): Origination Indication (ORIGIND)

(7): International

(152): Termination Restriction Code (TERMRES)

(2): Unrestricted

(48): CDMA Service Option List (IS-737) (CDMALIST)

9F812F0200029F812F0202019F812F020101

The MDN (the subscriber's number which shows in the calls or SMS is then set by the HLR (exactly as in GSM). The test which was done with Globalstar has a small particularity: the MDN Mobile Directory Number (MSISDN in GSM) is the same as the MIN Mobile Identity Number (IMSI in GSM). In general it is not the case with other CDMA networks. Also the CDMA MDN does not include the country code (+1).

In [6, figure 6] (i.e. Part I of this article) showing the number continuity GSM->Globalstar CDMA, the system will set the MDN sent to the Globalstar VLR = the GSM MSISDN. So when a call or SMS is made with the Globalstar, the GSM number will show.

3.3 Incoming Call to CDMA Subscriber

A call is made to the MSISDN of the GSM. The GSM HLR will send a PROVIDE ROAMING NUMBER (which includes the GSM IMSI) to the

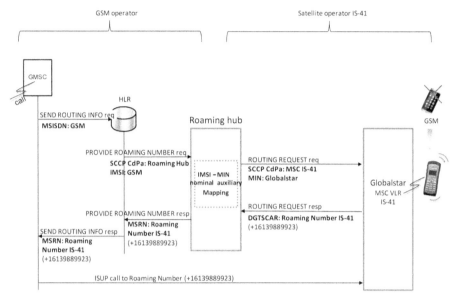

Figure 3 Incoming call to CDMA subscriber.

visited VLR which is the Roaming Hub. The Roaming Hub will map the
IMSI to the MIN and will send a ROUTING REQUEST including the MIN
(same as the IMSI in GSM). This ROUTING REQUEST also has the GT of
the GMSC (which is the GT of the Roaming Hub).

3.3.1 ROUTING REQUEST Request (Roaming Hub->VLR)

IS-41 Message Decoding © HALYS 2011, 2012
(16): ROUTING REQUEST (HLR->VLR Provide Roaming Number GSM)
Length = 52
    (129): Billing ID (BID)
        1E81C918587200
    (136): Mobile Identification Number MIN (as IMSI in GSM)
        MIN = +4037990012
    (137): Electronic Serial Number ESN (as IMEI in GSM)
        Manufacturer code = B3
        ESN = B309AFAF
    (149): MSC ID
        7809-201
    (150): System My Type Code (MYTYPE)
        (16): QUALCOMM

(47): CDMA Service Options (IS-737) (CDMASO)
    0002
(94): MSC Identification Number (as GT of GMSC in GSM) (MSCIN)
    Type of digit 00
    Nature of number 31
    International
    Presentation allowed
    Number is not available
    (2): Telephony Numbering E164
        (1): BCD
        Number of digits 11
        +33XXXXXXXXX /* GT of Roaming Hub */

3.3.2 ROUTING REQUEST Response (VLR->Roaming Hub)

IS-41 Message Decoding © HALYS 2011, 2012
(16): ROUTING REQUEST (HLR-¿VLR Provide Roaming Number GSM)
Length = 37
    (149): MSC ID
        7809-10
    (129): Billing ID (BID)
        1E810A65CF3C00
    (132): Digits (called MSISDN from GMSC or MSRN returned by VLR
in GSM) (DGTSCAR)
        Type of digit 06
        Nature of number 01
        International
        Presentation allowed
        Number is not available
        (2): Telephony Numbering E164
            (1): BCD
            Number of digits 11
            +16139889923 /* Roaming Number allocated by the VLR IS-41
            and returned to the Roaming Hub */
    (32): PC_SSN
        PC (24 bits) = 2247429 SSN = 7

The Roaming Hub will give +16139889923 in the GSM PROVIDE ROAM-
ING NUMBER Confirmation. As a result the GSM GMSC will call
+16139889923 directly and the IS-41 network will receive the same Mobile
Terminated Calls charges.

3.3.3 Call Forwarding IS-41 for Unsuccessful Mobile Terminated Calls

This does not work like GSM. In GSM the VLR profile contains "conditional call forwarding" informations for call busy, no response, not reachable. There is no such thing in IS-41, the profile returned by the HLR in the REGISTER NOTIFICATION result returned does not have it. When the incoming call of Figure 3 fails, the VLR IS-41 sends to the HLR a TRANSFER TO NUMBER REQUEST with the "Redirection Reason", asking for instructions.

IS-41 Message Decoding © HALYS 2011, 2012
(23): TRANSFER TO NUMBER REQUEST (VLR->HLR->VLR) (the VLR tells the result of a MT call and receives a redirection number)
Length = 22
    (136): Mobile Identification Number MIN (as IMSI in GSM)
        MIN = +4037990012
    (137): Electronic Serial Number ESN (as IMEI in GSM)
        Manufacturer code = B3
        ESN = B309AFAF
    (150): System My Type Code (MYTYPE)
        (16): QUALCOMM
    (147): Redirection Reason
        (4): No Page Response

The HLR responds by sending a "Redirecting Number" which could be the GSM VMS number, which is then called by the VLR. For the number continuity service, the Roaming Hub has extracted the GSM conditional call forwarding numbers from the INSERT SIBSCRIBER DATA and uses them to create the TRANSFER TO NUMBER Response sent to the VLR.

IS-41 Message Decoding © HALYS 2011, 2012
(23): TRANSFER TO NUMBER REQUEST (VLR->HLR->VLR) (the VLR tells the result of a MT call and receives a redirection number)
Length = 36
    (132): Digits (called MSISDN from GMSC or MSRN returned by VLR in GSM) (DGTSCAR)
        Type of digit 01
        Nature of number 00
        National
        Presentation allowed
        Number is not available
        (2): Telephony Numbering E164
            (1): BCD
            Number of digits 10

> +4037990012
(96): No Answer Time
> 0 seconds
(100): Redirecting Number Digits (number where call is forwarded, e.g. VMS)
> Type of digit 01
> Nature of number 01
> International
> Presentation allowed
> Number is not available
> (2): Telephony Numbering E164
>> (1): BCD
>> Number of digits 11
>> +33609000123 /* GSM VMS number */
(122): Termination Triggers
> (0): BUSY: Busy
> (4): RF: Failed call
> (8): NPR: No Page Response Call
> (12): NR: Member not reachable

The VLR will then forward the voice call to +33609000123 as shown in Figure 3.

3.4 Deregistration

This occurs when a subscriber powers down the handset. A signaling message is sent from the VLR to the HLR to deregister the handset.

IS-41 Message Decoding © HALYS 2011, 2012
(22): MS INACTIVE (VLR->HLR MS Purge GSM)
Length = 16
> (136): Mobile Identification Number MIN (as IMSI in GSM)
>> MIN = +4037990012
> (137): Electronic Serial Number ESN (as IMEI in GSM)
>> Manufacturer code = B3
>> ESN = B309AFAF

3.5 Incoming SMS-MT to CDMA Subscriber

3.5.1 The Local or Foreign SMSC Asks the HLR for the Visited MSC and MIN

This is the case of a SMSC, not the "number continuity" case which is simpler and does not need to interrogate the HLR IS-41.

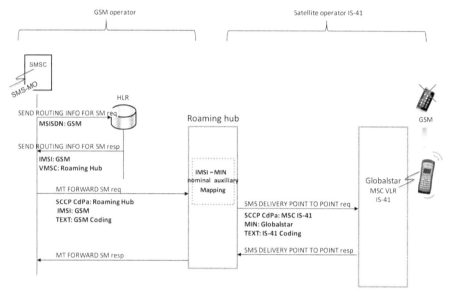

Figure 4 Incoming SMS-MT to CDMA subscriber.

The local or foreign SMSC uses the known public number MDN of the subscriber same as the MSIDN in a SRI_FOR_SM and gets the Visited MSC GT and the MIN (same as IMSI) which will be used in the SMS DELIVERY POINT TO POINT (same as FWD_SM_MT in GSM).

IS-41 Message Decoding © HALYS 2011, 2012
(55): SMS REQUEST (SMSC<->HLR Send Routing Information for SM in GSM)
Length = 24
    (109): SMS Notification indicator (demand of a HLR alert by the SMSC)
    (SMSNOTIND)
        Notify when available (01)
    (116): SMS Teleservice Identifier (IS-637) (SMSTID)
        CDMA Number Messaging Teleservice (1002)
(93): Mobile Directory Number MDN (as MSISDN for the SRI_FOR_SM or in
the INSERT SUBSCRIBER DATA GSM)
        Type of digit 05
        Nature of number 31
        International
        Presentation allowed
        Number is not available
        (2): Telephony Numbering E164
            (1): BCD

Number of digits 10
MDN = +4037990012

This is the response of the HLR including the MIN. This is particular (Globalstar) and they use a MDN (the MSISDN in GSM) equal to the MIN (the IMSI in GSM).

IS-41 Message Decoding © HALYS 2011, 2012
(55): SMS REQUEST (SMSC<->HLR Send Routing Information for SM in GSM)
Length = 29
    (137): Electronic Serial Number ESN (as IMEI in GSM)
        Manufacturer code = B3
        ESN = B309AFAF
    (104): SMS Address (as Visited MSC GT in GSM)
        Type of digit 05
        Nature of number 31
        International
        Presentation allowed
        Number is not available
        (1): ISDN Numbering Plan
            (2): IA5 International Alphabet 5
            Number of digits 11
            +16139889998
    (136): Mobile Identification Number MIN (as IMSI in GSM)
        MIN = +4037990012

3.6 The SMSC Sends the SMS to the Visited MSC

For number continuity the Roaming Hub does not need to interrogate the HLR IS-41 because it already knows the MIN and the visited VLR IS-41. It will send the SMS DELIVERY POINT TO POINT directly.

IS-41 Message Decoding © HALYS 2011, 2012
(53): SMS DELIVERY POINT TO POINT (SMSC->MSC SMS-MO or
MSC->SMSC SMS-MT Forward Short Message GSM)
Length = 188
    (136): Mobile Identification Number MIN (as IMSI in GSM)
        MIN = +4037990012
    (137): Electronic Serial Number ESN (as IMEI in GSM)
        Manufacturer code = B3
        ESN = B309AFAF
    (105): SMS Bearer Data
        MESSAGE_ID:

message type = 01 (Deliver (mobile terminated only))

message_ID: 0AD7

USER_DATA

Subparam length = 16

Msg_Encoding = 02 (7 bits ASCII)

Num 7b characters = 16

User_Data = Globalstar

NUMBER OF MSGs IN VMS: 12

LANGUAGE_INDICATOR: 02 (French)

MESSAGE CENTER TIMESTAMP: year = 2011 month = 12 day = 10 hour = 6 min = 45 sec = 32

VALIDITY PERIOD (absolute format): year = 2011 month = 12 day = 14 hour = 10 min = 0 sec = 0

VALIDITY PERIOD (relative format) = 85 (that is 25800 seconds)

ALERT ON MESSAGE DELIVERY: Use high priority alert

MESSAGE DISPLAY MODE (1): Mobile default setting: as predefined in the MS

REPLY OPTION:

User Ack (if this is SMS-MT) = Positive (manual) User ACK requested from the recipient user

Delivery Ack (if this is SMS-M0) = No Delivery ACK requested from the recipient

PRIORITY INDICATOR (2): Urgent

PRIVACY INDICATOR (3): Secret

DEFERRED DELIVERY TIME (absolute format): year = 2012 month = 1 day = 6 hour = 23 min = 59 sec = 0

DEFERRED DELIVERY TIME (relative format) = 84 (that is 25500 seconds)

USER RESPONSE CODE (predefined by SMSC for the SMSack) = 33

CALL BACK NUMBER

Digit Mode = 00 (DTMF (4 bits BCD))

Num_Fields = 13

Call Back Number = 1234567890ABC

CALL BACK NUMBER

Digit Mode = 01 (ASCII (8 bits))

Numbering type = 01

Numbering plan = 02

Num_Fields = 6

Call Back Number = 1234AB

(116): SMS Teleservice Identifier (IS-637) (SMSTID)

CDMA Voice mail notification (4099)

(109): SMS Notification indicator (demand of a HLR alert by
the SMSC) (SMSNOTIND)
Notify when available (01)

3.6.1 The HLR Alerts the SMSC When Subscriber Becomes Reachable

This is like ALERT SERVICE CENTRE in GSM. The SMSC will retry.

IS-41 Message Decoding © HALYS 2011, 2012
(54): SMS NOTIFICATION (HLR->SMSC Alert SC GSM)
Length = 29
    (137): Electronic Serial Number ESN (as IMEI in GSM)
        Manufacturer code = B3
        ESN = B309AFAF
    (136): Mobile Identification Number MIN (as IMSI in GSM)
        MIN = +4037990012
    (104): SMS Address (as Visited MSC GT in GSM)
        Type of digit 05
        Nature of number 31
        International
        Presentation allowed
        Number is not available
        (1): ISDN Numbering Plan
            (2): IA5 International Alphabet 5
            Number of digits 11
            +16139889998

3.7 Internet Data Services for CDMA->GSM Number Continuity

The likely usage is a GSM usage in a VPLMN. The Local Break-Out provides a simple solution so that the CDMA subscriber can have the internet access while using a GSM handset. But the GTP Protocol is common for GSM and CDMA2000, so that the PDP Context can also be established with the HPLMN CDMA GGSN [5].

4 CDMA<->GSM Number Continuity Service

This is a real case with major networks, such as SPRINT(CDMA) in the USA which provide the international roaming services to their subscribers with a dual standard handset CDMA plus a GSM SIM card from a "sponsor" and a Roaming Hub supplier providing the conversion. This is not a new idea

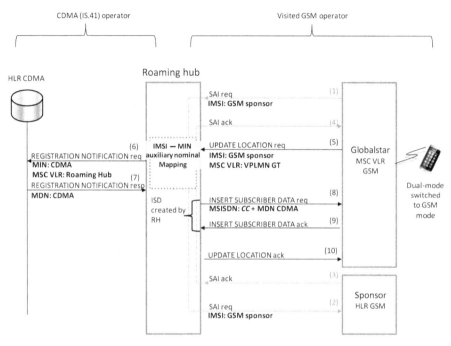

Figure 5 CDMA<->GSM registration with a Roaming Hub.

and the service was offered since 2003 with a separate GSM handset provide by a sponsor. It is more practical with dual standard handsets including the latest Iphone. When he arrives in Europe for example, the user selects the GSM mode or a use a separate GSM handset, the number continuity service is provided with the IS-41 HLR thanks to the two-way conversion in the Roaming Hub as shown by Figure 5. Compare with figures 5 and 6 of Part I [6].

In IS-41 there is no equivalent of MAP GSM INSERT SUBSCRIBER DATA, the MDN of the user (equivalent of the MSISDN) is sent by the HLR CDMA in the REGISTRATION NOTIFICATION resp. (7). The Roaming Hub creates an INSERT SUBSCRIBER DATA req (8) which contains the MSISDN to be used in GSM roaming. Figure 5 shows that

$$MSISDN = CC + MDNCDMA$$

CC would be +1 in the case of a US roamer, which is added by the Roaming Hub because in many cases the MDN does not include the country code CC.

References

[1] TIA/EIA-41-D. Number Radiotelecommunications Intersystem Operations, 1997. (Main description of the IS-41 protocol with the tables of operation and parameter codes.)

[2] TIA/EIA-637-A. Short Message Service for Spread Spectrum Systems, 1999. (Complement to [10.1] for the SMS service. Gives all the details and tables to implement the SMS service.)

[3] TIA/EIA-737. (Describes additional parameters to [1].)

[4] TIA/EIA-707-D. Based Network Enhancements for CDMA Packet Data Service (C-PDS). 3GPP2 N.S0029 V1.0.0, June 2002.

[5] IFAST#24/2004.10.04/07. CDMA Packet Data Roaming eXchange Guidelines.

[6] A. Henry-Labordère. The number continuity service: Part I, GSM<->satellite phone. Journal of Cyber Security and Mobility, 1(4):349–376, 2013.

Biography

Arnaud Henry-Labordère is a graduate engineer from Ecole Centrale de Paris (1966), Ph.D. (Mathematics, USA, 1968). He was professor of Operations Research at Ecole Nationale des Ponts et Chaussées during 25 years, as well as at Ecole Nationale des Mines de Paris. He is currently Visiting Professor at Prism-CNRS. He started at IBM research (1967) and founded three companies: FERMA (voice mail systems in 1983), Nilcom (first SMS network in 1999) and currently Halys (telecom equipments). He is the author of eight books (six in maths, two in telecoms) and has been granted 85 patents.

Online Manuscript Submission

The link for submission is: www.riverpublishers.com/journal

Authors and reviewers can easily set up an account and log in to submit or review papers.

Submission formats for manuscripts: LaTeX, Word, WordPerfect, RTF, TXT.
Submission formats for figures: EPS, TIFF, GIF, JPEG, PPT and Postscript.

LaTeX

For submission in LaTeX, River Publishers has developed a River stylefile, which can be downloaded from http://riverpublishers.com/river_publishers/authors.php

Guidelines for Manuscripts

Please use the Authors' Guidelines for the preparation of manuscripts, which can be downloaded from http://riverpublishers.com/river_publishers/authors.php

In case of difficulties while submitting or other inquiries, please get in touch with us by clicking CONTACT on the journal's site or sending an e-mail to: info@riverpublishers.com

www.ingramcontent.com/pod-product-compliance
Lightning Source LLC
LaVergne TN
LVHW012332060326
832902LV00011B/1850